Zephyr RTOS Cookbook

Build portable and scalable embedded systems through hands-on recipes

Dr. Roy Jamil
Dr. Ayoub Bourjilat

<packt>

Zephyr RTOS Cookbook

Portfolio Director: Rohit Rajkumar
Relationship Lead: Tanisha Mehrotra
Project Manager: Sandip Tadge
Content Engineer: Rohan Datta
Technical Editor: Tejas Vijay Mhasvekar
Indexer: Hemangini Bari
Production Designers: Aparna Bhagat
Growth Lead: Namita Velgekar

Production reference: 2210726

Published by Packt Publishing Ltd.
Grosvenor House
11 St Paul's Square
Birmingham
B3 1RB, UK.

ISBN 978-1-80742-917-1
www.packtpub.com

I would like to dedicate this book to my wife, whose patience and support carried me through every late night this book required, and to the Zephyr community: the developers, maintainers, and engineers who make work like this possible.

– Dr. Roy Jamil

To my wife, whose patience, support, and belief in me made this work possible.

– Dr. Ayoub Bourjilat

Foreword

Zephyr can be both impressive and intimidating at first glance. It is not just a small kernel or a collection of APIs; it is a full ecosystem for building modern embedded products, from board support and drivers to connectivity, security, testing, and long-term maintenance.

That is what makes a cookbook such a useful way to learn it. Most developers do not understand Zephyr by reading it front to back as an abstract architecture. They learn it by solving real problems: enabling a peripheral, describing hardware with devicetree, configuring a feature with Kconfig, writing a driver, debugging a thread, or bringing up a board.

Zephyr RTOS Cookbook meets developers in that practical space. It turns the breadth of Zephyr into approachable recipes, showing not only what to do, but why the pieces fit together the way they do. With each example, the system becomes less mysterious and more useful.

For anyone building firmware today, Zephyr offers a path toward portability, reuse, and collaboration without giving up the control embedded development requires. This book is a helpful companion on that path, whether you are taking your first steps with Zephyr or looking for patterns you can return to in production work.

The best way to learn Zephyr is to build with it. This book gives you a strong place to start.

Anas Nashif

Zephyr Project Technical Steering Committee Chair

Contributors

About the authors

Dr. Roy Jamil is an Embedded Systems Engineer, trainer, and consultant at AC6 specializing in Zephyr RTOS and embedded Linux. He holds a PhD in Embedded Systems focused on asymmetric multiprocessing platforms combining Linux and real-time operating systems. A Zephyr Project Technical Steering Committee (TSC) representative, Roy has delivered Zephyr professional training worldwide since 2022. He supports companies adopting Zephyr and porting new drivers and boards.

Dr. Ayoub Bourjilat is an Embedded Systems Engineer, trainer, and consultant at AC6. He holds a PhD in Embedded Systems and brings hands-on experience across RTOSes, processor architectures, and FPGA platforms. Through training and consulting, he focuses on robust system architecture, scalable Zephyr integration, and building auditable, secure-by-design embedded products aligned with modern industrial and IoT constraints.

We would like to thank Tanisha Mehrotra for giving us the opportunity to write this book. Our special thanks go to the reviewers, Erwan Gouriou, David Leach, and Rafiuddin, for their valuable input and for the time they spent reviewing this book, and to Anas Nashif for contributing the foreword. We would also like to thank Sandip Tadge, Rohan Datta, and the entire Packt team for their support throughout the writing of this book.

About the reviewer

Rafiuddin Syed is Director of Engineering at Drako Motors, where he leads the development of software-defined vehicle architectures and centralised computing platforms built on real-time operating systems. With over 20 years of experience in embedded systems, firmware engineering, RTOS-based platforms, and safety-critical automotive software, he has held senior roles at Intel, NVIDIA, Harman International, and Texas Instruments. His work spans the full embedded stack, including BSP development, device drivers, hardware abstraction layers, deterministic scheduling, power management, and connectivity subsystems. He has also contributed drivers to the mainline Linux and Android kernels and remains committed to advancing embedded computing, open-source RTOS ecosystems, and intelligent connected transportation.

Table of Contents

Preface

Zephyr is an open-source real-time operating system (RTOS) hosted by the Linux Foundation and designed for building connected, resource-constrained embedded devices. Beyond providing a deterministic kernel, Zephyr integrates device drivers, protocol stacks, system services, and development tools into a single platform. This lets developers build portable, scalable embedded applications that run across many hardware platforms while relying on consistent APIs and infrastructure.

Adopting Zephyr can feel very different from working with bare-metal systems, traditional RTOSes, or embedded Linux. Instead of tightly coupled, board-specific code, Zephyr follows a platform-oriented approach built around three ideas:

- Reusable components and standardized subsystems, rather than custom solutions for common tasks
- Hardware described separately from application logic, using devicetree
- Behavior driven by configuration, using Kconfig, rather than conditional code

This book treats Zephyr as a complete platform, not just a kernel. It is a recipe-led guide, and each recipe focuses on a practical task you will encounter when building Zephyr-based systems: structuring projects, configuring features, describing hardware, integrating drivers and subsystems, and reasoning about initialization, concurrency, and permissions at runtime.

You can follow the recipes in order to build a full mental model of Zephyr, or jump directly to the recipe that matches the task in front of you. By the end of the book, you will be able to develop maintainable Zephyr applications, adapt them to different hardware configurations, and scale your codebase across projects and teams.

Who this book is for

This book is for embedded developers and engineers who are transitioning to Zephyr from traditional RTOSes, bare-metal environments, or embedded Linux. It is aimed at readers with experience in embedded C who want to adopt Zephyr as a scalable platform for building portable systems that can be deployed consistently across multiple boards and development teams.

The target audience of this book is as follows:

- Embedded developers moving to Zephyr: They already write firmware in C and want a practical, task-oriented path into Zephyr's workspace model, configuration system (Kconfig and devicetree), and hardware abstractions.
- Engineers building portable, scalable products: They work on industrial or IoT devices that must run across several boards, and they need code that stays maintainable as the product and the team grow.
- Developers and architects evaluating Zephyr: They want to understand how West, Kconfig, devicetree, subsystems, and User Mode fit together before committing Zephyr to a real project.

No prior experience with Zephyr is required.

What this book covers

Chapter 1, Introducing the Zephyr Project, provides an introduction to the Zephyr development workflow. You will set up the environment, build your first application from the command line, and using VS Code with Workbench for Zephyr, flash and debug it on real hardware, explore the Zephyr source tree and ecosystem, and learn how to navigate the documentation.

Chapter 2, Structuring and Building Zephyr Applications, provides an overview of how Zephyr projects are organized and built. You will set up a West workspace, create and structure an application, understand the build process, and write a West manifest to define a workspace and organize projects.

Chapter 3, Configuring Zephyr Applications with Kconfig, provides an overview of how features are selected for a build. You will control features using configuration symbols, generate the merged configuration file, add custom symbols, control source compilation, edit configuration using menuconfig and guiconfig, and create persistent configuration.

Chapter 4, Using Devicetree to Describe Hardware, provides an overview of how Zephyr describes hardware at build time. You will inspect devicetree, create overlays, customize nodes and properties, use devicetree in application code, and write portable code with aliases.

Chapter 5, Exploring Zephyr's Device Driver Model, provides an overview of how applications interact with hardware through drivers. You will access hardware through the device model, configure and use GPIOs, configure and use GPIO interrupts, and write a custom devicetree binding.

Chapter 6, System Initialization and Multithreading, provides an overview of how Zephyr executes code concurrently. You will run code automatically at startup, create threads, delay and block threads, exchange data between threads, synchronize threads, and defer work using workqueues.

Chapter 7, Working with Zephyr Subsystems, provides an overview of the reusable subsystems that simplify application development. You will generate random numbers using the Random subsystem, configure and manage logging, interact with your application using the shell, broadcast messages using zbus, and trace and visualize runtime behavior.

Chapter 8, User Mode in Zephyr, provides an overview of how to run code with reduced privileges. You will enable User Mode, isolate global variables with memory domains, design an application around User Mode, and access kernel objects from user threads.

To get the most out of this book

You will need a working knowledge of the C programming language and an understanding of the basics of embedded development, including compiling firmware and flashing it onto a development board. No prior experience with Zephyr is required.

Software/Hardware covered in the book	OS Requirements
Zephyr RTOS and the SDK	Windows 10/11, Linux (Ubuntu 24.04 or newer), or macOS on Apple Silicon
West (the Zephyr meta-tool)	
Visual Studio Code with the Workbench for Zephyr extension	
Recommended: NXP FRDM-MCXN236 or ST NUCLEO-H563ZI development boards	
Recommended: OpenSmart Rich Shield	

The recipes are written for real hardware, but most can also be run in a simulation environment. For the development board, we recommend the NXP FRDM-MCXN236 or the ST NUCLEO-H563ZI, which are used throughout the book. Any other officially supported dev kit also works; you can follow most recipes on it directly, while a few will need minor board-specific adjustments.

Several recipes use external peripherals. We recommend the OpenSmart Rich Shield, which bundles many of the components used throughout the book. It can be obtained from various online retailers, for example:

`https://aliexpress.com/item/32840665547.html`

If you do not have the shield, you can follow those recipes using discrete components instead. At a minimum, you will need an LED and a potentiometer.

Download the example code files

This book includes a complete downloadable code bundle containing all the example projects and files used throughout the chapters. We recommend downloading the bundle so you can follow along smoothly and experiment with the examples.

Use the bundle as a practical starting point. Modify it, extend it, and apply what you learn by creating your own variations as you progress through the chapters.

Get the code bundle

If you bought the book directly from Packt:

1. Go to `packtpub.com`
2. Click your **profile picture** and select **Your Orders**
3. Find this book and click **Download Code**

If you bought this book from Amazon or any other channel partner:

1. Go to `packtpub.com/unlock` or scan the following QR code:

2. Search for this book
3. Sign up or log in to your free Packt account
4. Upload your proof of purchase and download the code bundle locally

> Note
>
> **Usage note:** You're free to use and modify this code for personal learning and non-commercial projects.

Download the color images

Your purchase includes a color, DRM-free PDF copy of this book, ideal for viewing color images, screenshots, and diagrams. Refer to Free benefits with your book section at the end of the Preface to unlock your PDF copy.

Conventions used

There are a number of text conventions used throughout this book.

`Code in text`: Indicates code words in text, database table names, folder names, filenames, file extensions, pathnames, dummy URLs, user input, and Twitter handles. Here is an example: "A West workspace is a structured collection of Git repositories coordinated by a manifest file named `west.yml`."

A block of code is set as follows:

```
#include <zephyr/kernel.h>

int main(void)
{
    printk("Hello from Zephyr\n");
    return 0;
}
```

Bold: Indicates a new term, an important word, or words that you see onscreen. For example, words in menus or dialog boxes appear in the text like this. Here is an example: "Zephyr development typically begins by creating a workspace and fetching the Zephyr source code using **West**, the meta-tool used to manage Zephyr repositories and orchestrate the build process."

Note

Warnings or important notes appear like this.

Tip

Tips and tricks appear like this.

Sections

In this book, you will find several headings that appear frequently (*Getting ready*, *How to do it...*, *How it works...*, *There's more...*, and *See also*).

To give clear instructions on how to complete a recipe, use these sections as follows:

Getting ready

This section tells you what to expect in the recipe and describes how to set up any software or any preliminary settings required for the recipe.

How to do it...

This section contains the steps required to follow the recipe.

How it works...

This section usually consists of a detailed explanation of what happened in the previous section.

There's more...

This section consists of additional information about the recipe in order to make you more knowledgeable about the recipe.

See also

This section provides helpful links to other useful information for the recipe.

Get in touch

Feedback from our readers is always welcome.

General feedback: If you have questions about any aspect of this book, mention the book title in the subject of your message and email us at `customercare@packtpub.com`.

Errata: Although we have taken every care to ensure the accuracy of our content, mistakes do happen. If you have found a mistake in this book, we would be grateful if you would report this to us. Please visit `www.packtpub.com/support/errata`, selecting your book, clicking on the Errata Submission Form link, and entering the details.

Piracy: If you come across any illegal copies of our works in any form on the Internet, we would be grateful if you would provide us with the location address or website name. Please contact us at `copyright@packt.com` with a link to the material.

If you are interested in becoming an author: If there is a topic that you have expertise in and you are interested in either writing or contributing to a book, please visit `authors.packtpub.com`.

Free benefits with your book

This book includes free benefits designed to support your learning and help you apply what you learn effectively. Activate them now for instant access (see the How to unlock section for instructions).

Here's a quick overview of what you can instantly unlock with your purchase:

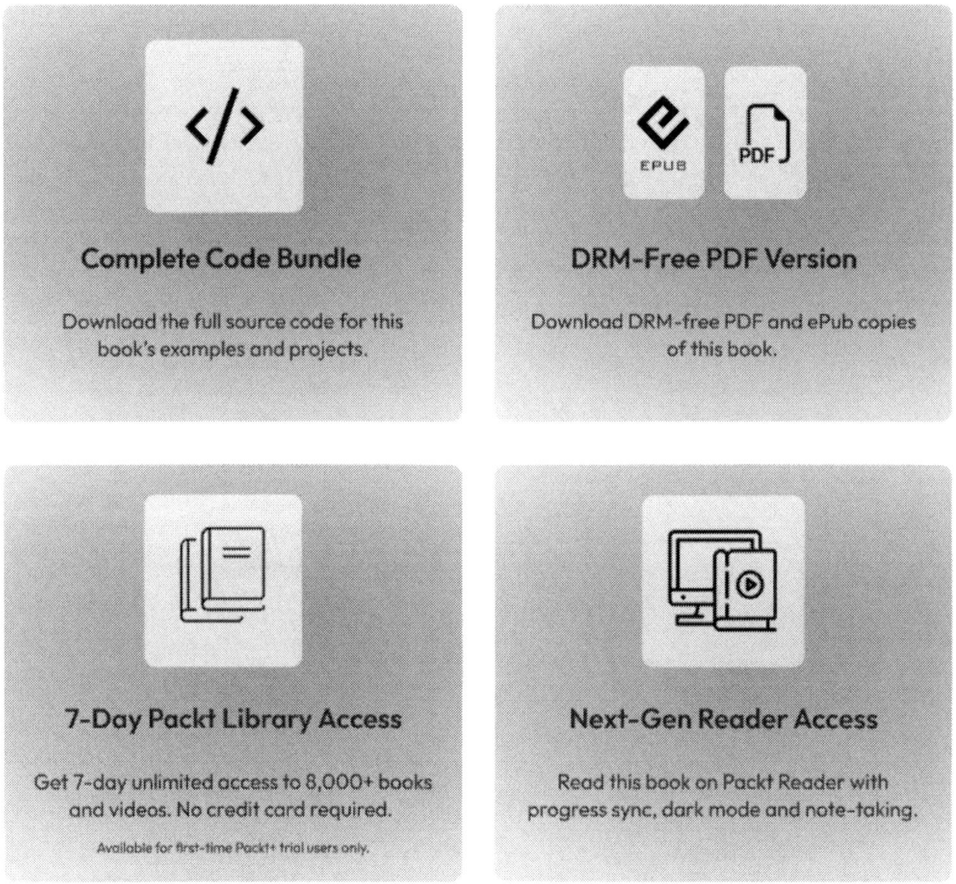

How to unlock

Scan the QR code (or go to `packtpub.com/unlock`). Search for this book by name, confirm the edition, and then follow the steps on the page.

UNLOCK NOW

Note: Have your invoice handy. Purchases made directly from the Packt website don't require an invoice.

Claim Your Free NXP Development Board

To help you get hands-on with Zephyr, NXP is offering a **free FRDM-MCXN236 development board*** to readers who redeem the coupon below. This offer is valid until available quantities are exhausted. The full list of its features is available on the `Zephyr documentation page`.

How to claim your free board

Visit the FRDM-MCXN236 product page: `FRDM-MCXN236 Development Board using MCUXpresso`

1. Add the board to your cart.
2. Create an NXP account (or sign in) and enter your shipping details.
3. Enter coupon code: **AMKRC9RG**
4. Once applied, the coupon makes both the board and shipping completely free.

Conditions

- This offer is valid until available quantities are exhausted.
- One board may be claimed per book purchase.
- Available only in countries and regions served by NXP's online store and shipping network. Delivery is subject to NXP's standard shipping coverage and any local import or customs requirements.

Questions?

For help redeeming your coupon or questions about the board, contact NXP at `support@nxp.com` or visit `https://www.nxp.com/support`.

**This giveaway is an independent effort by NXP Semiconductors. It is not managed by, affiliated with, or endorsed by Packt Publishing or the book's authors, none of whom bears any responsibility for the offer or its fulfillment.*

Share your thoughts

Once you've read *Zephyr RTOS Cookbook*, we'd love to hear your thoughts! Scan the QR code below to go straight to the Amazon review page for this book and share your feedback.

https://packt.link/r/1807429172

Your review is important to us and the tech community and will help us make sure we're delivering excellent quality content.

1

Introducing the Zephyr Project

The **Zephyr Project** is an open-source **real-time operating system (RTOS)** designed for building connected, resource-constrained embedded devices. Beyond providing a deterministic kernel, Zephyr integrates drivers, protocol stacks, system services, and development tools into a single platform. This approach allows developers to build portable and scalable embedded applications that can run across many hardware platforms while relying on consistent APIs and infrastructure.

Getting started with a new embedded platform involves several steps, including setting up the development environment, building and running applications, and understanding how the platform is organized. In this chapter, you will walk through the core Zephyr development workflow by setting up the environment, building your first application, running it on real hardware, and exploring the structure of the Zephyr ecosystem. These steps provide the foundation needed to understand the configuration, hardware description, and subsystem concepts that will be explored in later chapters.

In this chapter, we will cover the following recipes:

- Building your first Zephyr application from the command line
- Building your first Zephyr application using VS Code
- Flashing and debugging Zephyr applications
- Exploring the Zephyr source tree and ecosystem
- Navigating the Zephyr documentation

Note

Your purchase includes a free PDF copy + code bundle

Your purchase includes a DRM-free PDF copy of this book, the code bundle, and additional exclusive extras. See the *Free benefits with your book* section in the *Preface* to unlock them instantly and maximize your learning.

Technical requirements

To follow the recipes in this chapter, you will need:

- A supported host machine running one of the following operating systems:
    - Windows 10/11 on Intel/AMD x86-64
    - Linux on Intel/AMD x86-64, preferably Ubuntu 24.04 or newer
    - macOS on Apple Silicon (ARM-based)
- Other Linux distributions are also supported. Refer to the Zephyr documentation for the complete list of supported host environments.
- A supported development board or a simulation environment.
- Preferably one of the following reference boards used throughout the book:
    - `NXP FRDM-MCXN236`
    - `ST NUCLEO-H563ZI`

Any other officially supported board can also be used. You only need to select the matching board name and adapt any board-specific settings later in the book.

The code files for this chapter are available in the downloadable code bundle under the Chapter 1 folder. You can download the code by following the instructions in the **Download the example code files** section in the *Preface*.

Building your first Zephyr application from the command line

Zephyr development typically begins by creating a workspace and fetching the Zephyr source code using **West**, the meta-tool used to manage Zephyr repositories and orchestrate the build process.

In this recipe, you will install the required host tools, download the **Zephyr SDK**, create a Python virtual environment, initialize a Zephyr workspace, and build the Hello World sample from the command line.

> **Note**
>
> This recipe demonstrates the command-line workflow. The next recipe shows how to perform the same steps using Workbench for Zephyr, which provides a graphical environment built on top of the same tools and commands. For the rest of the book, it is recommended to use Workbench for Zephyr while still understanding the underlying command-line workflow executed in the background.

Getting ready

Zephyr requires several host tools in order to build applications. These include:

- **CMake**
- **Ninja**
- **Python**
- **Git**
- **Devicetree compiler**
- Additional utilities used during configuration

On Windows, the easiest way to install these dependencies is by using winget.

Open a PowerShell terminal and run the following command to install the required host tools:

```
winget install Kitware.CMake Ninja-build.Ninja oss-winget.gperf Python.Python.3.12
Git.Git oss-winget.dtc wget 7zip.7zip
```

This installs the most common host tools required for Zephyr development.

For the most up-to-date installation instructions and platform-specific requirements, refer to the official Zephyr documentation: `https://docs.zephyrproject.org/latest/develop/getting_started/index.html`

> **Note**
>
> These manual steps are useful when you want to configure the environment yourself. The next recipe demonstrates an alternative approach using Workbench for Zephyr.

How to do it...

1. **Create a workspace directory**

   In PowerShell, create a workspace directory and move into it:

```
mkdir zephyr-workspace
cd zephyr-workspace
```

2. **Create a Python virtual environment**

   Create a Python virtual environment:

```
python -m venv .venv
```

   Activate it in PowerShell:

```
.\.venv\Scripts\Activate.ps1
```

   Using a virtual environment keeps Zephyr Python dependencies isolated from the rest of the system.

3. **Install West**

   Install the Zephyr meta-tool west inside the virtual environment:

```
pip install west
```

   West is used to:

   - manage the Zephyr workspace
   - download repositories
   - build applications
   - flash and debug firmware

4. **Initialize the workspace and fetch Zephyr**

   Create the workspace using `west init`, then fetch Zephyr and its modules:

```
west init zephyr-workspace --mr v4.4.1
cd zephyr-workspace
west update
```

   At this point, the workspace contains the Zephyr repository together with the additional modules required by the project.

Zephyr is not a single repository. Instead, it is composed of multiple coordinated repositories that together form the Zephyr ecosystem.

5. **Install Zephyr Python dependencies**
   Once the repositories have been downloaded, install the additional Python packages required by Zephyr:

```
west packages pip --install
```

6. **Install the Zephyr SDK**
   The **Zephyr SDK** contains the toolchains for Zephyr-supported architectures, together with additional host tools used during development.

```
west sdk install
```

West downloads and installs the Zephyr SDK for you.

> **Note**
>
> If needed, west sdk install also supports options to choose a different installation destination or to install only selected toolchains. You can see the available options with:
>
> `west sdk install --help`

Alternatively, it is also possible to install the SDK manually from the official release page: `https://github.com/zephyrproject-rtos/sdk-ng/releases`

The easiest option is to download the full SDK archive. If you target only a specific architecture, you can also construct a smaller SDK installation by extracting only the required toolchain. For example, for **ARM Cortex-M** boards, extracting only the arm-zephyr-eabi toolchain is often sufficient.

7. **Build the Hello World sample**
   To verify that the development environment works, build the Hello World sample for the `FRDM-MCXN236` board:

```
west build -p always -b frdm_mcxn236 samples\hello_world
```

The -p always option forces a pristine build. This is recommended for new users because it ensures the build directory is recreated cleanly.

For building the same sample for the NUCLEO-H563ZI board or any other board, place the output in a different build directory:

```
west build -p always -b nucleo_h563zi samples\hello_world -d build_nucleo
```

Using separate build directories allows both board builds to coexist without overwriting each other.

8. **Verify the result**
   After a successful build, the firmware image for the first build can be found in:

```
build\zephyr\zephyr.elf
```

At this point, the Zephyr development environment has been successfully validated.

How it works...

Building a Zephyr application combines several inputs:

- the application source code
- the Zephyr platform (kernel, drivers, subsystems)
- the Zephyr SDK toolchain

These inputs are processed by the build system to generate the final firmware image.

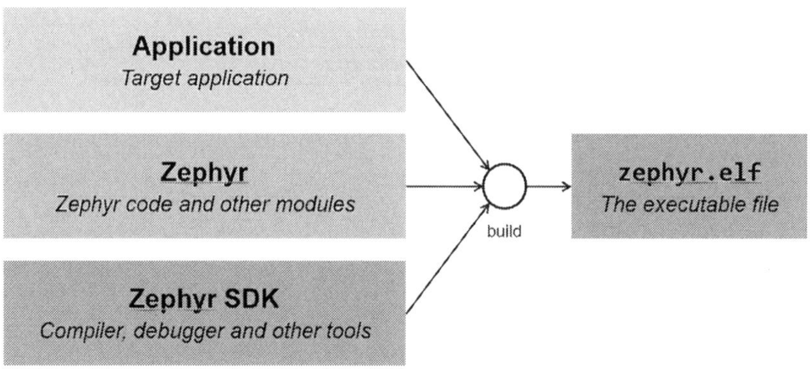

Figure 1.1 – Zephyr build flow from sources to firmware image

The diagram illustrates how the target application, the Zephyr codebase, and the SDK toolchain are combined during the build process to produce the final executable file.

Most configuration decisions in Zephyr are resolved at build time, including enabled drivers, hardware configuration, and system features. This approach produces a firmware image tailored to the selected board and application.

The build process also relies on several environment variables that define where different components are located:

- `ZEPHYR_BASE`
- `ZEPHYR_SDK_INSTALL_DIR`
- `EXTRA_ZEPHYR_MODULES`

These variables allow the build system to locate the Zephyr codebase, SDK toolchain, and optional modules.

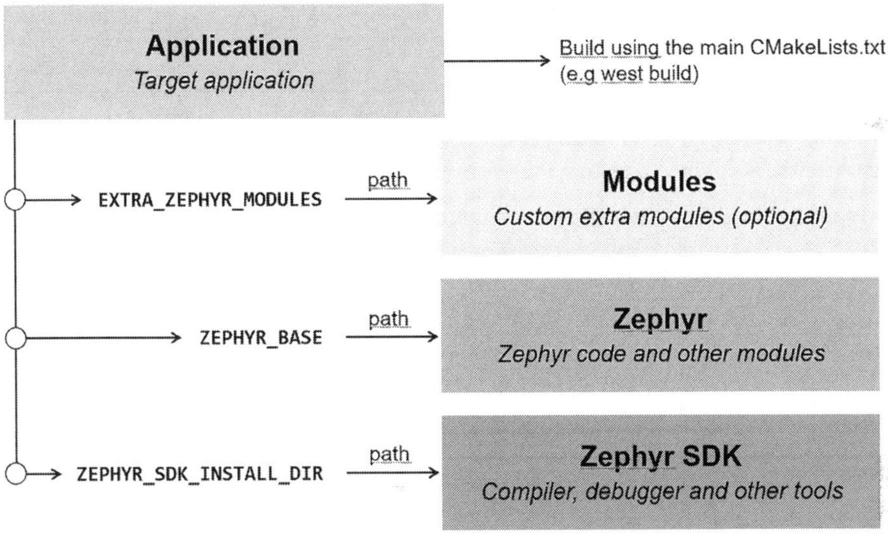

Figure 1.2 – Workflow of the Zephyr build system

There's more...

In this example, the Hello World sample was built directly from the Zephyr source tree:

```
west build -p always -b frdm_mcxn236 samples\hello_world
```

This is convenient for validating the environment and experimenting with samples.

In real projects, applications are typically developed outside the Zephyr repository. Keeping application code separate from the Zephyr source tree avoids modifying Zephyr internals and simplifies upgrades.

A common workflow is to copy a sample as a starting point:

```
cp -r zephyr/samples/hello_world my_app
```

The new application can then be built with:

```
west build -p always -b frdm_mcxn236 my_app
```

In larger projects, applications and external modules are often managed using a custom west manifest. This allows the workspace layout and participating repositories to be controlled explicitly. Custom manifests and workspace organization will be discussed later in this book.

In addition, while west is the standard and recommended tool for Zephyr development, it is technically possible to build Zephyr applications without west.

In that case, the build can be driven directly using CMake, but the developer must manually:

- clone the Zephyr repository
- clone additional modules
- configure the `ZEPHYR_MODULES` variable
- manage repository dependencies

While this approach works, it is significantly more complex and therefore rarely used in practice.

For this reason, most Zephyr workflows rely on west to manage the workspace and coordinate the build process.

See also

- Zephyr Project documentation – Getting Started Guide: `https://docs.zephyrproject.org/latest/develop/getting_started/index.html`
- Zephyr Project documentation – West (Zephyr meta-tool): `https://docs.zephyrproject.org/latest/develop/west/index.html`
- Zephyr SDK releases: `https://github.com/zephyrproject-rtos/sdk-ng/releases`
- Practical Zephyr - Zephyr Basics (Part 1): `https://interrupt.memfault.com/blog/practical_zephyr_basics`

Building your first Zephyr application using VS Code

While Zephyr development can be performed entirely from the command line, many developers prefer a graphical environment that simplifies workspace management, toolchain installation, and project creation.

Workbench for Zephyr is a Visual Studio Code extension that provides an integrated environment for Zephyr development. It simplifies several tasks that otherwise require multiple command-line steps, such as installing host tools, importing toolchains, creating west workspaces, and generating new applications.

In this recipe, you will install the Workbench for Zephyr extension, configure the required host tools and toolchain, create a Zephyr workspace, and build a Hello World application. This workflow relies on the same underlying tools used in the previous recipe, including west, CMake, and the Zephyr SDK, but Workbench automates much of the configuration.

Getting ready

You will need to install Visual Studio Code if it is not installed yet. Open Visual Studio Code and navigate to the Extensions view. Search for and install the extension: Workbench for Zephyr.

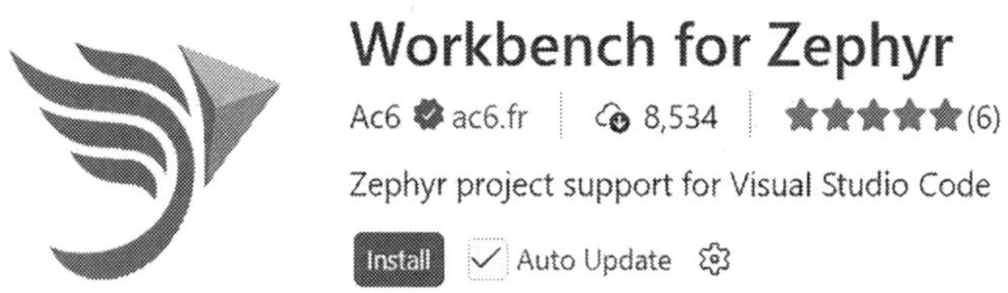

Figure 1.3 – Workbench for Zephyr

How to do it...

1. **Install the required host tools**

   After installing the extension, open the Workbench for Zephyr panel in the left sidebar.

   Locate the Host Tools section and click **Install Host Tools**

Workbench installs the required host dependencies automatically, including tools such as CMake, **Ninja**, Python support utilities, and other required components.

*Figure 1.4 – **Install Host Tools** panel*

2. **Import the Zephyr SDK toolchain**
   Next, import the Zephyr SDK toolchain.

   In the Workbench for Zephyr panel, locate the Toolchains section and select **Import Toolchain**.

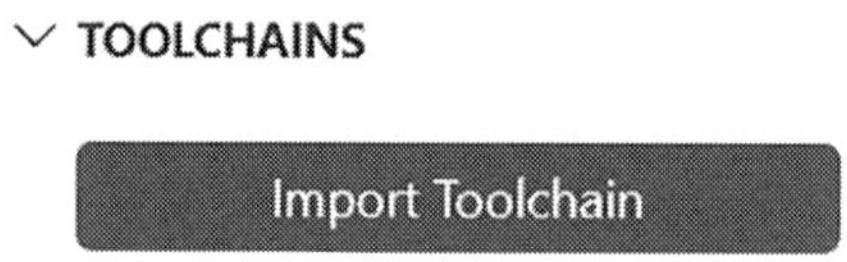

Figure 1.5 – Toolchains panel

In the **Add Toolchain** dialog:

- Select Toolchain family: **Zephyr SDK**
- Select Source: **Official**
- Select SDK Type: **Minimal**
- Select Version: **v1.0.1**
- Select the required architecture: **arm**
- Choose the installation location and click **Import**.

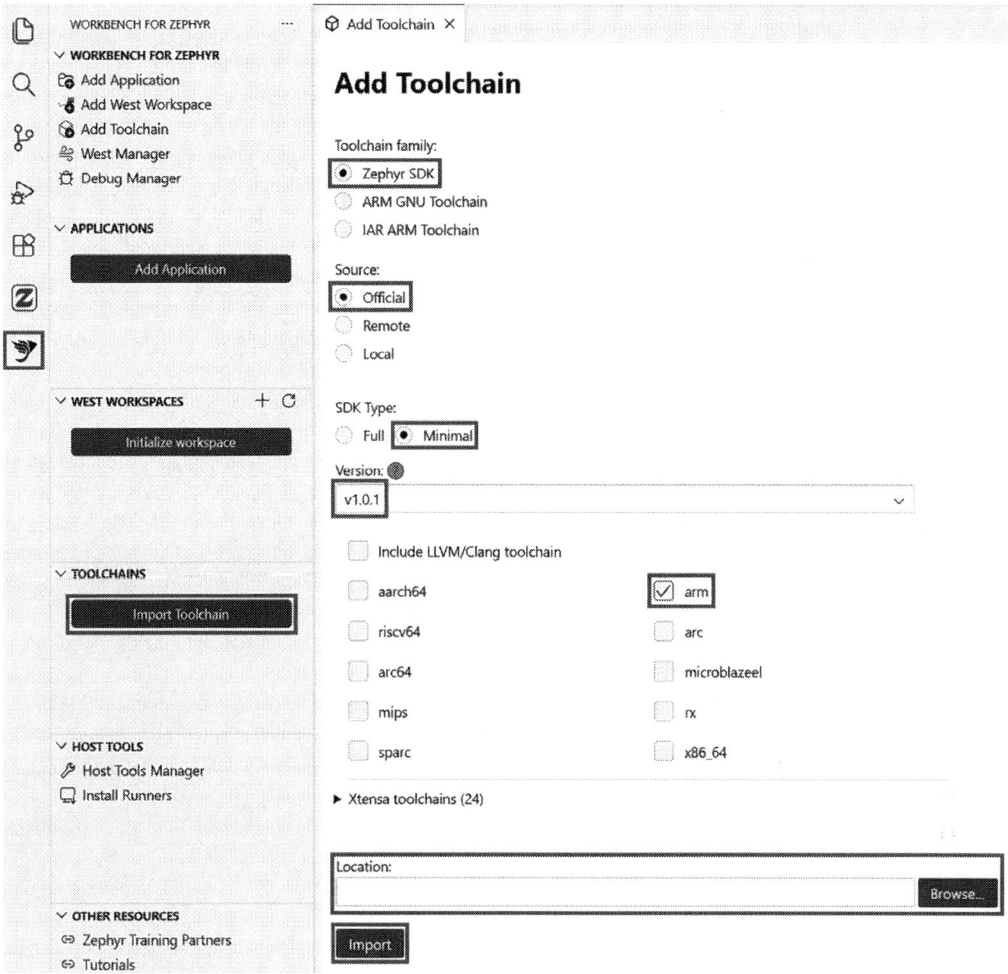

Figure 1.6 – Add a toolchain using Workbench for Zephyr

> **Note**
>
> A minimal SDK installation is sufficient for this book because both reference boards are ARM-based. If you later target other architectures, you can import a fuller SDK installation or add more architectures.
>
> You can also optionally select the LLVM Clang toolchain during SDK installation (or add it later) and use it as an alternative to GCC when building Zephyr applications.

3. **Create a west workspace**

   Next, create a Zephyr workspace.

   In the Workbench for Zephyr panel, select **Initialize workspace**.

   In the Create west workspace dialog:

   - Choose Source location: Minimal **from template**
   - Select Template: **NXP** (or STM32 for STM32 boards)
   - Select Revision: **v4.4.1**
   - Choose the workspace location: It is recommended to use subfolder

   Click **Import** to create the workspace.

   Workbench will automatically:

   - Initialize the west workspace
   - Download Zephyr and its modules
   - Configure the workspace environment

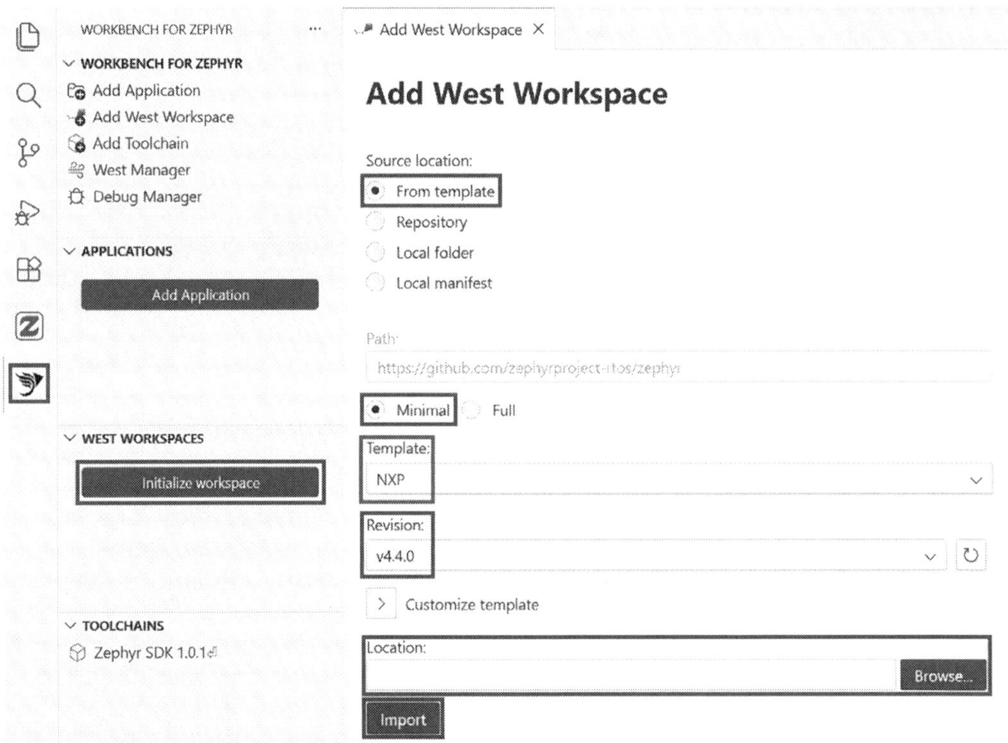

Figure 1.7 – Creating a west workspace with Workbench for Zephyr

> **Note**
>
> You can expand Customize template and add both the NXP and STM32 modules to make your workspace compatible with both vendors' boards. Alternatively, select Full to import all default modules.

4. **Create a new Zephyr application**

   Once the workspace has been created, generate a new application project.

   In the Applications section of the Zephyr Workbench panel, select: **Create New Application**

   Configure the project:

   - Select the **west workspace**
   - Select the Zephyr SDK toolchain version 1.0.1
   - Select the board, for example NXP FRDM-MCXN236
   - Select Sample project: **hello_world**
   - Enter a project name: keep the default or rename the project
   - Select the **project location**
   - Click Create.

   Workbench will generate the project and configure the build environment automatically.

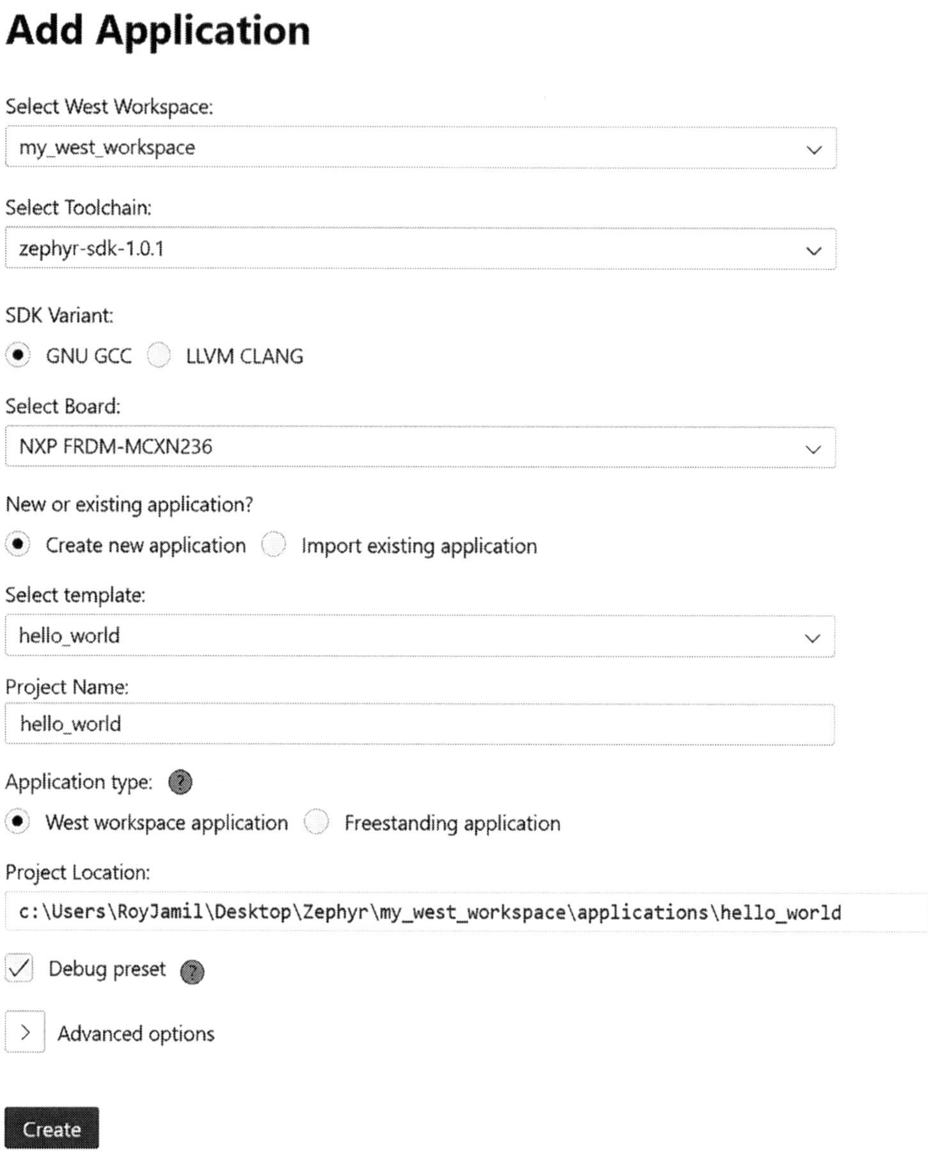

Figure 1.8 – Creating a new Zephyr application in Workbench for Zephyr

> **Note**
>
> You can repeat the same process later with the ST NUCLEO-H563ZI board to confirm that the same sample can be generated for a second target.

5. **Build the application**

   Once the **application** has been created, build the project using the Workbench interface, right-click on the application then click on **build**.

   After the build completes successfully, the compiled firmware will be available in the project build directory.

How it works...

Workbench for Zephyr is a graphical layer built on top of the Zephyr development tools.

Internally it performs the same steps shown in the previous recipe:

- Manages the west workspace
- Installs required host tools
- Configures the Zephyr SDK
- Generates application templates
- Invokes the Zephyr build system

Workbench simplifies these operations by providing graphical dialogs that configure the environment automatically.

When a project is created or built, Workbench executes the corresponding west commands in the background. This means the graphical workflow remains fully aligned with standard Zephyr development practices.

There's more...

Workbench for Zephyr provides several additional capabilities that can significantly simplify development and analysis tasks beyond simply building and running applications.

One useful feature is the contextual menu available when right-clicking an application project in the Workbench interface. From this menu, several development operations can be performed directly without using command-line tools.

For example, configuration of the Zephyr build can be performed using graphical tools:

- `Menuconfig`: a terminal-based configuration interface for **Kconfig** options
- `GUIConfig`: a graphical interface similar to `menuconfig`

These interfaces allow developers to explore and modify the configuration parameters defined in `prj.conf` and the broader Zephyr configuration system.

Workbench also integrates tools that help analyze memory usage and binary size. For example:

- Memory reports can be generated to inspect RAM and flash consumption
- PUNcover for analyzing memory usage and stack worst-case scenarios consumption

Workbench also integrates seamlessly with the **Devicetree** language server (DTS LSP), which provides advanced editing features for Devicetree files inside Visual Studio Code. These features include:

- syntax highlighting
- validation
- navigation between nodes and bindings

This significantly improves the experience when working with Devicetree overlays and board descriptions.

Another example is support for static analysis tools. Workbench can integrate with tools such as ECLAIR, which provide MISRA compliance checking and advanced static analysis capabilities. These tools help detect potential coding issues early in the development process and assist teams working under strict coding standards such as MISRA C.

Together, these features transform Workbench for Zephyr into a complete development environment that combines build management, configuration tools, code analysis, and debugging capabilities in a single interface.

See also

- YouTube – Getting Started with Zephyr Workbench in VS Code: `https://www.youtube.com/watch?v=1RB0GI6rJk0`
- Workbench for Zephyr documentation: `https://z-workbench.com/`
- Zephyr Project documentation – Workbench for Zephyr: `https://docs.zephyrproject.org/latest/develop/tools/workbench_for_zephyr.html`
- Workbench for Zephyr VS Code Marketplace: `https://marketplace.visualstudio.com/items?itemName=Ac6.zephyr-workbench`
- Puncover example: `https://z-workbench.com/docs/documentation/memory-analysis/puncover/`

Flashing and debugging Zephyr applications

After building an application, the next step in the development workflow is flashing the firmware onto the target board and debugging the program. Zephyr provides integrated flashing and debugging support through the west tool and a set of utilities known as runners.

Workbench for Zephyr simplifies this workflow by integrating these capabilities directly into Visual Studio Code. It automatically detects the build output, configures the debug environment, and launches the debugging session.

In this recipe, you will configure the required runners, flash the firmware to the target board, and start a debugging session using Workbench for Zephyr.

Behind the scenes, Workbench uses the same Zephyr debugging infrastructure used from the command line. It launches the appropriate debug runner and starts a **GDB** debug session connected to Visual Studio Code.

Getting ready

Make sure the Hello World application from the previous recipe builds successfully. Connect the development board to your computer using USB.

How to do it...

1. **Install the required runners**
   Zephyr uses small utilities called runners to communicate with the debug probe and flash firmware to the device.

   Workbench for Zephyr provides a runner manager that installs these tools automatically.

   Open the **Workbench** panel and navigate to:

   Host Tools Manager → **Install Runners**

   In this list you will see several available runner packages.

   For the boards used in this book install:

   - NXP **LinkServer** for Microcontrollers
   - STM32CubeCLT

LinkServer is used for debugging NXP boards, while STM32CubeCLT automatically installs **STM32CubeProgrammer** and STLink GDB server that are used for STM32 devices.

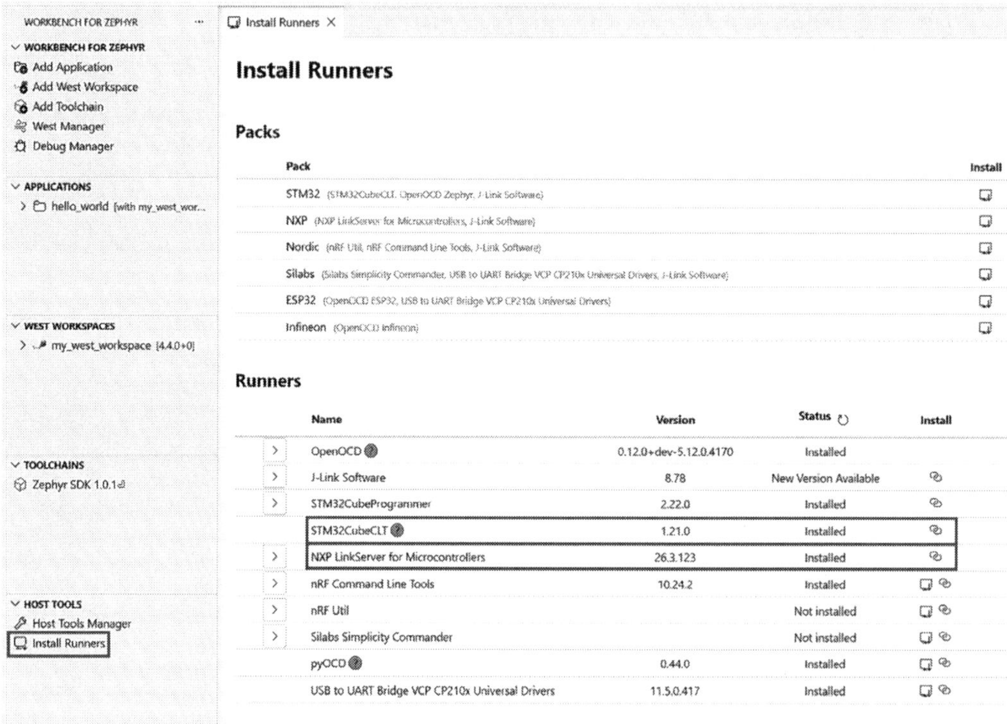

Figure 1.9 – Installing runners

2. **Flash the firmware**

   Once the runners are installed, flashing the firmware can be done directly from the application context menu.

   In the Workbench panel:

   1. Right-click on the application (**hello_world**)
   2. Select **Flash**

You should choose the appropriate runner for the selected board.

For example:

- NXP boards typically use the LinkServer runner
- STM32 boards typically use STM32CubeProgrammer

The firmware built earlier will then be programmed onto the board.

3. **Start a debugging session**
   To start debugging, open the Debug Manager in the Workbench panel.

   Select: **Debug Manager**

   In the Debug Manager window:

   1. Select the application **(hello_world)**
   2. Select the **build** configuration
   3. Select the runner corresponding to your board

   For the NXP FRDM board, choose LinkServer.

   Workbench will automatically detect the executable produced during the build:

```
${workspaceFolder}\build\primary\zephyr\zephyr.elf
```

   The debug components are also detected automatically.

   Click **Debug** to start the debugging session.

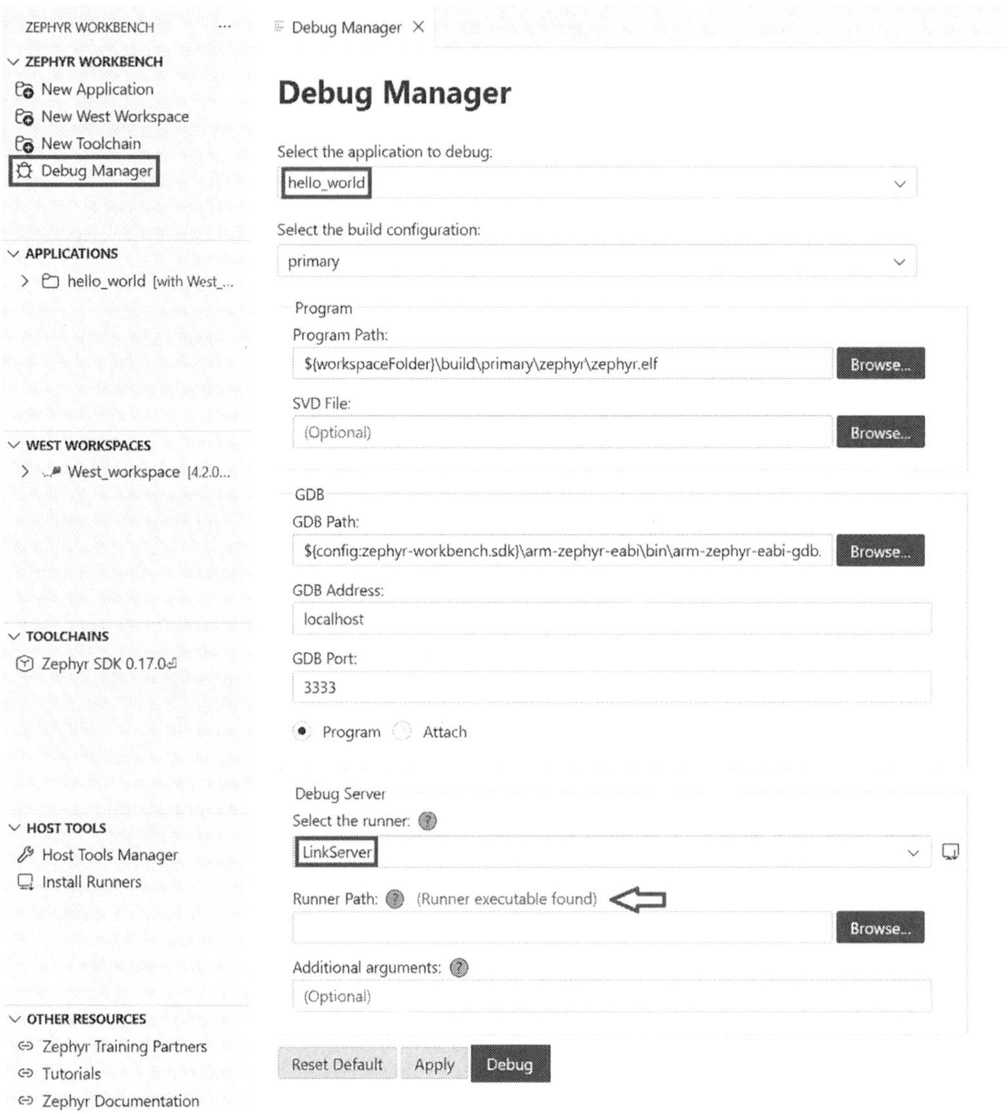

Figure 1.10 – Using the Debug Manager in Workbench for Zephyr

4. **(Optional) Provide the SVD file**

   For advanced debugging, you may optionally provide an SVD file.

   SVD files describe the registers and peripherals available on the microcontroller. When an SVD file is provided, the debugger can display peripheral register states during execution.

Workbench attempts to detect the correct SVD file automatically. If it cannot find one, the file can be selected manually in the Debug Manager using the SVD File field.

5. **Observe the debug session**
   Once debugging starts, Visual Studio Code connects to the target through GDB.

   You can now:

   - set breakpoints
   - step through code
   - inspect variables
   - view registers
   - monitor peripheral state

   The Serial Monitor panel can also be used to view program output from the UART console.

Figure 1.11 – A working debug session

How it works...

Zephyr flashing and debugging is based on the concept of runners.

A runner is a small program that knows how to interact with a specific debug probe or flashing tool. Examples include:

- LinkServer for NXP devices
- STM32CubeProgrammer for STM32 boards
- OpenOCD for many supported probes

When debugging starts, Workbench internally launches the appropriate runner. This runner starts a debug server connected to the target device.

At the same time, Workbench launches the Zephyr toolchain's GDB client and connects it to the debug server.

The connection between GDB and Visual Studio Code allows the IDE to control program execution, set breakpoints, and inspect memory and registers.

Internally, Workbench relies on the same Zephyr debugging infrastructure used by the command-line workflow. The extension orchestrates these tools automatically and connects them to the Visual Studio Code debugger.

There's more...

Zephyr applications can also be flashed and debugged directly from the command line using west.

For example, the following command starts a debug session:

```
west debug
```

This command launches the appropriate debug runner and starts the GDB debugging environment.

However, when working purely from the command line the developer must manually interact with the debugger, manage the terminal sessions, and configure the debug environment.

While this approach is fully supported and sometimes useful for advanced workflows, most developers prefer the integrated debugging experience provided by Workbench for Zephyr.

See also

- YouTube – Getting Started with Zephyr Workbench in VS Code: `https://www.youtube.com/watch?v=1RB0GI6rJk0`
- Workbench for Zephyr documentation: `https://z-workbench.com/`
- Zephyr Project documentation – Debugging: `https://docs.zephyrproject.org/latest/develop/debug/index.html`
- STM32CubeCLT: `https://www.st.com/en/development-tools/stm32cubeclt.html`
- LinkServer for Microcontrollers: `https://www.nxp.com/design/design-center/software/development-software/mcuxpresso-software-and-tools-/linkserver-for-microcontrollers:LINKERSERVER`

Exploring the Zephyr source tree and ecosystem

After building your first Zephyr application using Workbench for Zephyr, the next step is to understand how the Zephyr workspace and source tree are organized. Zephyr is more than a traditional RTOS kernel. It is a complete platform composed of drivers, subsystems, hardware support packages, and development infrastructure.

In this recipe, you will explore the Zephyr workspace created in the previous recipes and examine how the source tree is organized. You will also see how the application created earlier sits alongside the Zephyr platform sources.

Understanding this structure will help you navigate the Zephyr codebase, locate drivers and subsystems, and understand how applications interact with the platform.

Getting ready

This recipe assumes that you have completed the previous recipe using Workbench for Zephyr and that:

- Zephyr version 4.4.1 has been imported
- Zephyr SDK version 1.0.1 is installed
- A Hello World application has been created

Open Visual Studio Code with Workbench for Zephyr and load your workspace.

The exact workspace layout may vary depending on how the west manifest is configured. In this book, the Zephyr sources and modules are stored in a dedicated directory named "deps" defined by the manifest, while applications are stored alongside it in separate folders.

How to do it...

1. **Locate the workspace structure**
   Open the workspace folder in Visual Studio Code. In the Explorer view you should see a structure similar to the following:

```
workspace
|
├── deps
|   ├── zephyr
|   └── modules
|
└── hello_world
```

   The directory containing Zephyr and its modules is defined by the west manifest configuration. Applications are typically placed outside of the Zephyr repository so that they remain independent from the platform sources. The `hello_world` directory contains the application created in the previous recipe.

2. **Inspect key directories**

   The Zephyr repository contains the core Zephyr platform. Some of the most important directories include:

Directory	Description
arch	Architecture-specific implementations (ARM, RISC-V, x86, etc.)
boards	Board definitions and hardware descriptions
drivers	Hardware drivers for peripherals such as GPIO, UART, SPI, I²C, and sensors
include	Public header files used throughout the system
kernel	The Zephyr real-time kernel implementation
subsys	Higher-level services such as networking, logging, storage, and device management
samples	Example applications demonstrating Zephyr APIs
tests	Automated tests used to validate the platform

Table 1.1 - Overview of key directories in the Zephyr source tree

3. **Locate board support files**

   Zephyr supports many development boards. Board definitions are located inside the boards directory and are organized by vendor.

   For example, the boards used in this book can be found at:

   - zephyr/boards/nxp/frdm_mcxn236
   - zephyr/boards/st/nucleo_h563zi

   Each board directory contains files describing the hardware platform, including:

   - `Devicetree files` (.dts) describing peripherals and hardware resources
   - Default configuration files (_`defconfig`)
   - Build configuration files used by the Zephyr build system

   These files allow the same application to run on multiple boards with minimal changes.

4. **Explore the samples directory**

   Zephyr provides many example applications that demonstrate different platform capabilities.

   Open the `zephyr/samples` directory.

   ```
   Samples are grouped by functionality, for example:
   samples/basic
   samples/drivers
   samples/net
   samples/bluetooth
   samples/subsys
   ```

   These examples serve as reference implementations that demonstrate how to use Zephyr APIs and subsystems.

5. **Inspect the application created earlier**

   Return to the workspace root and open the application created in the previous recipe; the structure typically looks like this:

   ```
   hello_world
   │
   ├── src
   │   └── main.c
   ├── CMakeLists.txt
   └── prj.conf
   ```

   The `src` directory contains the application source code.

   `CMakeLists.txt` integrates the application with the Zephyr build system.

   `prj.conf` defines configuration options that enable or disable Zephyr features.

How it works...

When examining the Zephyr source tree, one important observation becomes clear: the kernel represents only a small portion of the overall codebase. The majority of the platform consists of drivers, subsystems, tests, and supporting infrastructure.

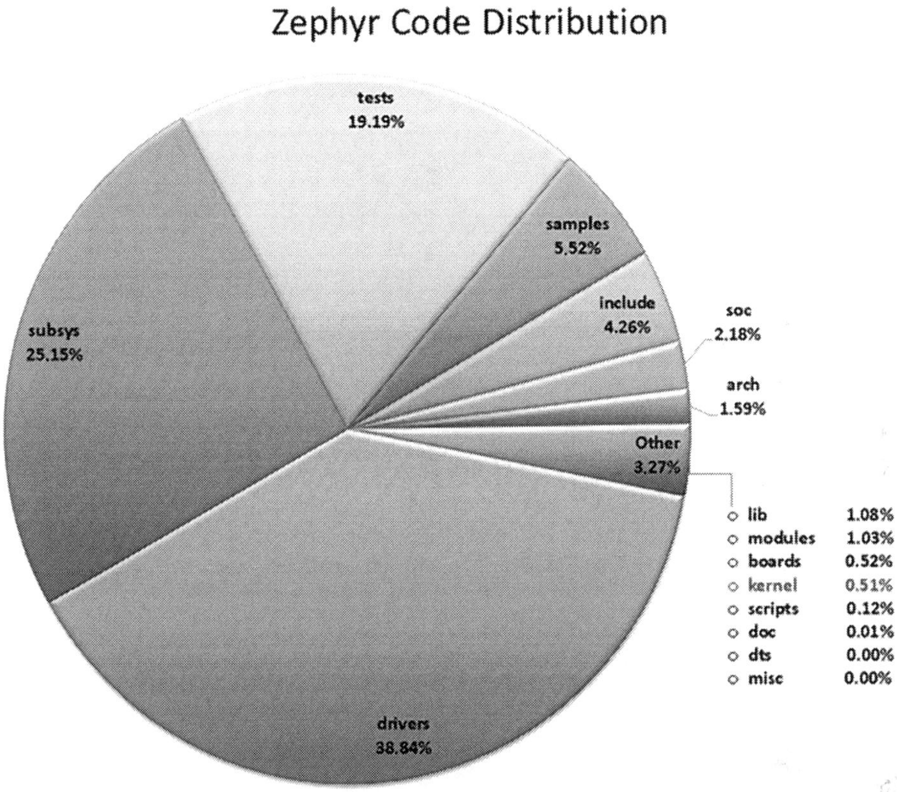

Figure 1.12 – Distribution of code across major Zephyr source directories, version 4.3.0

The figure illustrates how the Zephyr source code is distributed across the major directories of the project. Drivers represent the largest portion of the codebase, followed by subsystems and tests, while the kernel itself occupies only a small percentage.

This distribution reflects Zephyr's architectural philosophy. Zephyr is not intended to be only a small scheduling kernel. Instead, it is designed as a complete embedded platform where the kernel provides the foundation and most functionality is implemented through drivers and subsystems built on top of it.

Drivers implement standardized APIs for common hardware interfaces such as GPIO, UART, SPI, I^2C, and sensors. Subsystems build on top of these drivers to provide higher-level functionality including networking stacks, Bluetooth support, logging systems, storage services, and device management.

Because applications interact with these stable APIs rather than hardware-specific implementations, the same application can often run on multiple boards with minimal modifications.

There's more...

Zephyr is released under the Apache License, Version 2.0, a permissive open-source license widely used in industrial software.

In practice, this means Zephyr can be used in commercial or proprietary products without requiring developers to publish their application source code. Applications built on top of Zephyr can therefore remain fully proprietary.

If you modify Zephyr's own source files, the license requires that those modifications be clearly indicated for traceability. Typically, this means adding a short comment near the top of the modified file describing the changes.

Unlike copyleft licenses, the Apache License does not require applications built on top of Zephyr to publish their source code.

The Apache License also includes an explicit patent grant from contributors, which provides additional legal protection when using the software.

As with most open-source licenses, the software is provided without warranty or liability. The license also does not grant permission to use Zephyr trademarks or project branding in products.

See also

- Zephyr Project GitHub repository: `https://github.com/zephyrproject-rtos/zephyr`
- Zephyr Project documentation – Architecture overview: `https://docs.zephyrproject.org/latest/introduction/index.html`

Navigating the Zephyr documentation

Zephyr documentation is an integral part of the project and evolves together with the source code. It is written and reviewed by the same engineers who contribute to the platform, which means it focuses on practical development rather than high-level descriptions. As a result, the documentation often provides both conceptual explanations and concrete implementation details.

However, the Zephyr documentation should not be treated as a single linear book. Instead, it is a collection of different resources designed to answer different types of questions. Learning how to navigate these resources efficiently is an important skill when working with Zephyr.

In this recipe, you will learn how the documentation is structured and how to quickly find the information needed during development, whether it is understanding a feature, locating an API definition, finding a working example, or identifying configuration options.

Getting ready

Open the Zephyr documentation website in your browser: `https://docs.zephyrproject.org`.

Make sure you are viewing the documentation for the same version used in your workspace. In this book the examples are based on Zephyr version 4.4.1.

The documentation interface includes a navigation sidebar and a search bar that allow you to quickly move between guides, references, and examples.

How to do it...

1. **Find out how a Zephyr feature works**
   When learning about a Zephyr feature or subsystem, the best starting point is the main documentation guides.

   These sections explain the purpose of a feature, how it integrates with the system, and how it should be used in applications. They often include small code snippets that demonstrate correct usage patterns.

   For example, topics such as kernel services, connectivity, and system subsystems are explained in the main documentation sections.

   `https://docs.zephyrproject.org`

2. **Verify the exact API**
   Once you understand the concept, you may need to check the exact API definition.

   The API reference is generated using Doxygen directly from the Zephyr source code. It contains precise information such as function prototypes, data structures, parameters, return values, and constraints.

   This reference is particularly useful when writing code and verifying the exact behavior of an API.

   `https://docs.zephyrproject.org/latest/doxygen/html/`

3. **Find a working example**
   When learning a feature, a complete example is often the fastest way to understand how everything fits together.

The samples and demos section provides many small applications that demonstrate how to use specific Zephyr features. Each sample usually includes a short description, build instructions, and expected output.

`https://docs.zephyrproject.org/latest/samples/index.html`

Another useful source of examples is the Zephyr test suite. Tests are designed to validate specific features and APIs, and they often contain minimal but realistic code demonstrating how a component is intended to be used. Unlike samples, tests are not typically listed in the main documentation, but they can be explored directly in the Zephyr source tree or on the project's GitHub repository.

4. **Check board capabilities**
   Board documentation describes the hardware supported by a specific development board.

   Board pages typically include information such as supported peripherals, console configuration, flashing instructions, and debugging notes.

   This is often the fastest way to answer questions such as whether a board supports a particular feature or which interface is used for the console.

   `https://docs.zephyrproject.org/latest/boards/index.html`

5. **Find configuration options**
   Many Zephyr features are controlled through configuration options defined by Kconfig.

   The Kconfig reference lists configuration symbols together with their dependencies, default values, and intended purpose.

   This information is useful when editing the `prj.conf` file of an application.

   `https://docs.zephyrproject.org/latest/kconfig.html`

6. **Check Devicetree properties**
   Hardware configuration in Zephyr is described using Devicetree.

   Bindings define the schema for Devicetree nodes, including which properties exist, which are required, and what their values represent.

   When writing or modifying Devicetree nodes, the bindings reference is the authoritative source for valid properties and their meanings.

   `https://docs.zephyrproject.org/latest/build/dts/api/bindings.html`

How it works...

The Zephyr documentation is tightly connected to the source code. Many parts of the documentation are automatically generated from the project itself.

For example, the API reference is generated from source code comments, the Kconfig reference is generated from configuration definitions, and Devicetree bindings documentation is generated from hardware description schemas.

This approach ensures that the documentation stays synchronized with the codebase. As features evolve, the associated documentation evolves as well.

There's more...

The documentation website includes a built-in search function that allows you to quickly locate topics, APIs, configuration options, and board pages.

Another important aspect is that the Zephyr documentation itself is open source. It is stored in the Zephyr repository alongside the code, which means developers can propose improvements or corrections through pull requests.

If you want to contribute to the documentation or other parts of the project, the Zephyr contribution guidelines describe the workflow, coding standards, and review process:

`https://docs.zephyrproject.org/latest/contribute/guidelines.html`

See also

- Zephyr Discord community: `https://discord.com/invite/Ck7jw53nU2`
- Zephyr community resources: `https://www.zephyrproject.org/community`

Get this book's PDF copy, code bundle, and more

Scan the QR code (or go to `packtpub.com/unlock`). Search for this book by name, confirm the edition, and then follow the steps on the page.

UNLOCK NOW

Note: Have your invoice handy. Purchases made directly from the Packt website don't require an invoice.

2

Structuring and Building Zephyr Applications

A Zephyr application is more than a set of source files. It exists within a structured workspace, integrates with a multi-stage build system, and depends on multiple repositories working together. Without understanding this structure, it becomes difficult to build, scale, or maintain Zephyr-based projects effectively.

In practice, developers often need to set up a workspace, organize application code, and manage dependencies while ensuring that builds remain reproducible across environments. Misunderstanding how these pieces fit together can lead to build failures, inconsistent setups, or tightly coupled project structures that are hard to extend.

This chapter focuses on the practical aspects of structuring and building Zephyr applications. You will learn how to set up a West workspace, create and organize a Zephyr application, understand the build process, and define a workspace using a West manifest. The recipes also demonstrate how these components interact to produce a working firmware image.

Together, these recipes provide a structured approach to organizing Zephyr projects and building them in a consistent and scalable way.

In this chapter, we will cover the following recipes:

- Setting up a West workspace
- Creating and structuring a Zephyr application
- Building a Zephyr application
- Writing a West manifest and organizing projects

Technical requirements

To follow the recipes in this chapter, you will need:

- A supported host machine
- Zephyr SDK or a compatible toolchain
- Python environment required by Zephyr
- A supported development board (e.g., STM32, NXP) or a simulation environment
- Visual Studio Code with Workbench for Zephyr extension (optional but recommended)

The code files for this chapter are available in the downloadable code bundle under the Chapter 2 folder. You can download the code by following the instructions in the **Download the example code files** section in the *Preface*.

Setting up a West workspace

A Zephyr project depends on multiple repositories, including the kernel, modules, and application code. Managing these manually is complex and error-prone.

West provides a workspace model that organizes and synchronizes all required repositories. Without a properly initialized workspace, it is not possible to build or manage a Zephyr project correctly.

In this recipe, you will create a West workspace, populate it with the required repositories, and explore its structure.

Getting ready

Before starting, make sure you have a working Zephyr development environment.

You can use either:

- A command-line west environment
- Workbench for Zephyr properly installed

How to do it...

1. **Create a workspace directory**
   Create a new directory for your workspace:

   ```
   mkdir west-workspace
   cd west-workspace
   ```

2. **Initialize the West workspace**

Run the following command:

```
$ west init
```

This operation creates a workspace directory and prepares the internal metadata that West needs to manage it. From this point on, the directory tree is treated as a West workspace.

A workspace has a clearly defined root. The root is identified by the presence of a hidden directory named:

```
.west/
```

3. **Fetch the workspace repositories**
   Once the workspace exists, its repositories are populated using the command:

```
$ west update
```

This operation fetches and synchronizes all Git repositories that belong to the workspace.

> **Note**
>
> When using the Workbench for Zephyr extension for VS Code, these operations are typically performed automatically. Initializing and updating the workspace are handled by the extension behind the scenes, while still relying on the same West commands.

After the update is completed, inspect the workspace directory tree.

A minimal workspace might look like this:

```
west-workspace/
├── .west/
│   └── config
├── zephyr/
├── modules/
│   └── hal/
│       └── vendor_hal/
└── app/
    ├── src/
    ├── prj.conf
    └── CMakeLists.txt
```

To display all repositories managed by West, use the command:

```
west list
```

How it works...

Terminology:

Term	Meaning
Workspace	A folder where West clones and manages all your Zephyr-related code
Project	A Git repository managed by West (e.g., zephyr, hal_nxp, etc.)
Manifest	A file, `west.yml`, that tells West which repos you need and where to find them
Commands	Operations you run with West, like west build, west flash, etc.

Table 2.1 – Common West workspace terminology

A West workspace is a structured collection of Git repositories coordinated by a manifest file named `west.yml`.

The manifest defines:

- Which repositories belong to the workspace
- Where they are located
- Which revisions are checked out

Each repository listed in the manifest is called a project. A West workspace is identified by the presence of the `.west/` directory at its root. When a West command runs, it searches upward in the directory tree until it finds this directory, then performs all operations relative to that workspace root. Inside the workspace, multiple repositories coexist, including the Zephyr repository, which contains the kernel, subsystems, drivers, and build system. Applications are typically placed alongside the Zephyr repository rather than inside it, and a workspace may contain one or many applications.

Optional repositories may also be present, such as:

- Hardware abstraction layers provided by silicon vendors
- Bootloaders such as MCUboot
- Debugging or tracing tools
- Shared middleware libraries

These repositories remain independent Git projects. West does not merge or rewrite them; instead, it ensures that they are present, placed in known locations, and synchronized according to the manifest. This creates a structured workspace rather than a monolithic codebase while still allowing developers to work independently on applications, drivers, or the Zephyr kernel itself. One of the main benefits of this model is reproducibility: the manifest captures the complete set of repositories and revisions required for the project, allowing another developer to recreate the same workspace state simply by running `west init` and `west update`.

There's more...

West provides additional commands that operate across the entire workspace, including:

- `west flash`
- `west update`
- `west twister`
- `west forall`
- `west blobs`
- `west sign`
- `west spdx`

Advanced projects may use custom commands or multiple manifests to manage complex product structures.

See also

- Zephyr Project documentation – Getting Started Guide: `https://docs.zephyrproject.org/latest/develop/getting_started/index.html`
- West (Zephyr meta-tool) documentation: `https://docs.zephyrproject.org/latest/develop/west/index.html`

Creating and structuring a Zephyr application

A Zephyr application must follow a specific structure to be recognized by the build system. Without this structure, the application cannot be built or integrated with Zephyr.

In this recipe, you will create a minimal Zephyr application with the required files and directory layout and prepare it for integration into a Zephyr workspace.

Getting ready

Before starting, make sure you have a working Zephyr development environment, either using the command line or Workbench for Zephyr. You should also confirm that your Zephyr workspace is initialized and functional.

How to do it...

1. **Create the application directory**

   Create a new directory for your application:

   ```
   my_app/
   ```

   Inside it, create the following structure:

   ```
   my_app/
   ├── CMakeLists.txt
   ├── prj.conf
   └── src/
       └── main.c
   ```

   This is the minimal and canonical Zephyr application layout.

2. **Add the application source file**

   Create `src/main.c` with the following content:

   ```
   #include <zephyr/kernel.h>

   int main(void)
   {
       printk("Hello from Zephyr\n");
       return 0;
   }
   ```

   This file contains the application entry point.

   The `main()` callback is executed in the context of the main thread, which is created and started automatically by Zephyr during system startup.

3. **Add the configuration file**

   Create `prj.conf` in the application root.

   For now, this file is empty, but it is required to exist.

It serves to set application-specific Kconfig options.

> **Note**
>
> As an example, you could add:
>
> ```
> CONFIG_PRINTK=y
> ```
>
> In this case, this option is not required because it is enabled by default.

4. **Add the build description**
   Create `CMakeLists.txt` in the application root:

```
cmake_minimum_required(VERSION 3.20.0)

find_package(Zephyr REQUIRED HINTS $ENV{ZEPHYR_BASE})
project(my_app)

target_sources(app PRIVATE src/main.c)
```

This file tells the Zephyr build system which source files belong to the application.

5. **Build the application**
   From your workspace, run via command line:

```
west build -b <board> path/to/my_app
```

If you are using Workbench for Zephyr, right-click the application and select Build.

The build produces several outputs, including:

```
<build-dir>/zephyr/zephyr.elf
```

How it works...

A Zephyr application follows a predictable layout:

- `CMakeLists.txt` connects the project to Zephyr's build system
- `prj.conf` defines configuration options for this specific application
- Source files are placed inside the `src` directory
- Additional resources such as overlays or extra modules can be added later

This separation keeps application logic independent from Zephyr's internal implementation. The application supplies behavior, while Zephyr supplies the kernel, drivers, and subsystems.

All inputs are merged into a single executable image.

The application does not build Zephyr directly; instead, it is linked into the Zephyr build.

There's more...

As applications grow, you may add:

- DeviceTree overlays such as <board>.overlay
- Additional source directories
- Board-specific configuration fragments

Applications can live inside the Zephyr repository, inside a West workspace, or in standalone repositories.

These variations become important when working with West and manifests, which are covered later in this chapter.

See also

- Zephyr Application Development Guide: `https://docs.zephyrproject.org/latest/develop/application/index.html`
- Zephyr CMake build system: `https://docs.zephyrproject.org/latest/build/cmake/index.html`

Building a Zephyr application

A Zephyr application goes through multiple build stages before producing the final executable. When build issues occur, it is important to understand where in the process they happen. In this recipe, you will rebuild a Zephyr application and observe the different stages of the build system.

Getting ready

Reuse the my_app project created in the previous recipe. Ensure it builds successfully.

How to do it...

1. **Rebuild the application**
   Start a pristine build to clearly observe all configuration stages.

   A pristine build removes the existing build directory before starting a new build. This forces CMake, Kconfig, and Devicetree to run from scratch instead of reusing cached results.

   Using the command line, run:

   ```
   west build -b <your_board> path/to/my_app -p
   ```

   The -p option performs a pristine build.

   If you are using Workbench for Zephyr:

   Right-click the application and select **Rebuild (Pristine)**.

2. **Observe the build phases**
   During a pristine build, the build directory is cleaned and regenerated. This ensures that:

   - CMake configuration is re-executed
   - Kconfig is reprocessed
   - Devicetree is reprocessed
   - No cached configuration is reused

   Watch the console output carefully. You will typically see two major messages:

   ```
   -- west build: generating a build system
   ```

   Next, you will see:

   ```
   -- west build: building application
   ```

This corresponds to the actual build phase (usually executed by Ninja). During this stage:

- Source files are compiled
- The Zephyr kernel, drivers, and application are compiled
- All object files are linked into the final executable

3. **Inspect the generated outputs**
   After the build completes, inspect these outputs inside the build directory:

```
<build-dir>/
└── zephyr/
    ├── zephyr.elf
    ├── zephyr.hex
    ├── zephyr.bin                  (may exist depending on configuration)
    ├── zephyr.dts                  (merged Devicetree output)
    └── include/generated/zephyr/
        ├── autoconf.h              (Kconfig output)
        └── devicetree_generated.h  (Devicetree macros)
```

   These files are generated during different phases of the build and are regenerated automatically when the configuration changes.

How it works...

The Zephyr build process follows a structured sequence of responsibilities. The diagram below illustrates the high-level flow.

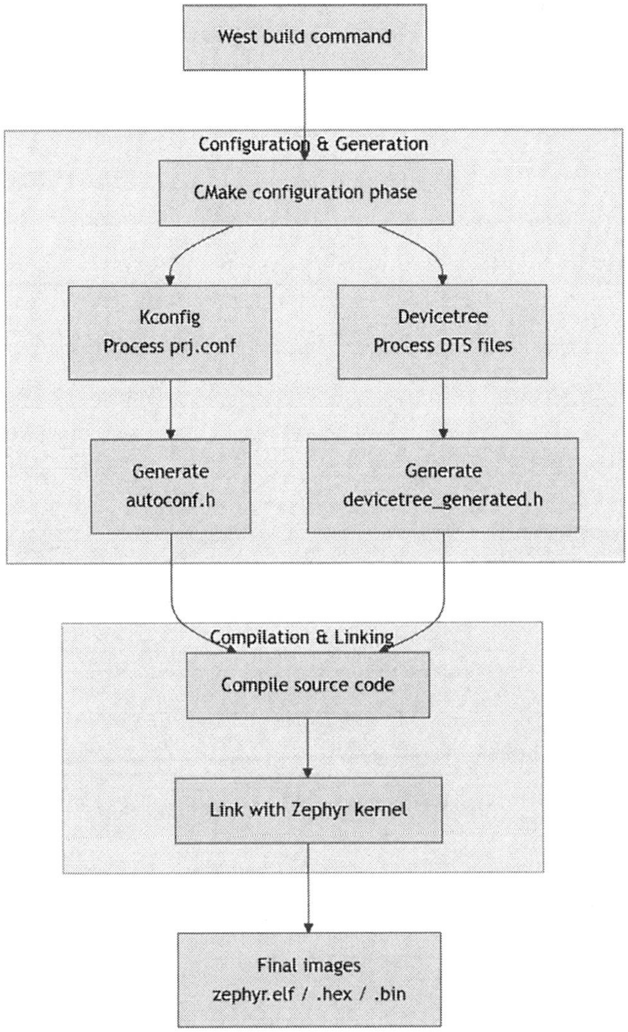

Figure 2.1 – Zephyr build process overview

The process begins when you run `west build`. West acts as an orchestrator. It selects the board, identifies the application, and invokes the underlying build system. West does not configure features or process hardware descriptions.

CMake then configures the build. It loads the Zephyr build system, processes all `CMakeLists.txt` files, determines which modules participate in the build, and triggers the configuration systems.

Two major configuration systems run during this phase:

- Kconfig: It determines which software features are enabled. Kconfig answers questions such as which subsystems are enabled and which drivers are selected.
- Devicetree: It describes the hardware. It defines which peripherals exist, their addresses, interrupts, and configuration properties.

After configuration is complete, compilation begins. The compiler builds:

- Application source files
- Zephyr kernel
- Enabled drivers
- Enabled subsystems

Finally, the linker combines all object files into the final executable image. The main output is `zephyr.elf`, which can be converted into other formats such as hex or bin for flashing.

All configuration decisions are resolved before compilation begins. The compiler and linker only process the results of earlier stages.

There's more...

A common mistake is trying to use the wrong mechanism for a task.

Examples of incorrect usage include:

- Using Kconfig to describe hardware pins
- Using devicetree to enable software features
- Treating West as a build system rather than an orchestrator

Keeping these roles clear will help you diagnose build problems and structure applications correctly as they grow.

See also

- Zephyr Build System documentation: `https://docs.zephyrproject.org/latest/build/index.html`

- Kconfig documentation in Zephyr: `https://docs.zephyrproject.org/latest/build/kconfig/index.html`
- Devicetree in Zephyr: `https://docs.zephyrproject.org/latest/build/dts/index.html`

Writing a West manifest and organizing projects

A Zephyr workspace is defined by a manifest file that specifies which repositories are used, where they are fetched from, and how they are organized locally. Without a properly defined manifest, the workspace cannot be reproduced, maintained, or shared reliably.

In addition, the way repositories are organized within the workspace impacts scalability and long-term maintainability. Poor organization can lead to unnecessary dependencies and complex project structures.

The West manifest is a YAML file named `west.yml`. It describes the repositories (called projects in West terminology) that belong to a workspace. The manifest may live in different locations depending on the chosen workspace topology, but its purpose remains the same: to define the composition of the workspace.

At a high level, the manifest answers four questions:

- Which repositories are part of this workspace?
- Where are they fetched from?
- Which revision of each repository is used?
- Where is each repository placed locally?

In this recipe, you will create a West manifest and learn how to organize repositories using recommended workspace topologies.

Getting ready

Before starting, make sure you have a working Zephyr development environment.

You can use either:

- A command-line west environment
- Workbench for Zephyr properly installed

How to do it...

1. **Create a minimal manifest file**
   Create an empty file named `west.yml`.

2. **Define the basic structure (optional)**

```
manifest:
  projects:
    - name: example-project
      url: https://example.com/example-project
      revision: main
```

   Each entry under the projects section defines one Git repository managed by West.

   A project entry commonly includes:

   - `name`: a unique identifier within the manifest
   - `url` or remote + repo-path: where the repository is fetched from
   - `revision`: the Git reference to check out
   - Path: the local directory where the repository is placed

3. **Define a remote**
   To avoid repeating long URLs, manifests commonly define remotes. A remote defines a base URL that can be reused by multiple projects. Replace the content of `west.yml` with:

```
manifest:
  remotes:
    - name: zephyrproject
      url-base: https://github.com/zephyrproject-rtos

  projects:
    - name: zephyr
      remote: zephyrproject
      repo-path: zephyr
      revision: v4.4.1
```

   In this case, the full repository URL is constructed from the remote's URL base and the project's repo path.

   Remotes improve readability and make it easier to change repository hosting locations later.

4. **Specify the local repository path**
   The **path** field defines where a repository is placed relative to the workspace root.

The manifest repository itself can also be configured using the **self** section.

This is commonly used when the manifest lives inside the application repository. Example:

```
manifest:
  self:
    path: my_app
```

This tells West where the manifest repository should be placed in the workspace.

Manifests can include projects defined in other manifests using import.

This is most commonly used to reuse Zephyr's own manifest, which already defines many optional modules.

For example:

```
import:
    path-prefix: deps
    name-allowlist:
        - cmsis_6
        - hal_nxp
        - hal_stm32
        - percepio
```

It instructs West to import projects from `zephyr/west.yml`, but only those explicitly listed in the allowlist, and to place them under the given path prefix.

This approach allows you to reuse existing manifests without pulling in unnecessary repositories.

5. **Fetch the manifest projects**
   Initialize the workspace using the local manifest repository:

   ```
   west init -l .
   ```

   Then fetch all projects defined in the manifest:

   ```
   west update
   ```

   West clones the repositories listed in `west.yml` and checks out the specified revisions.

In most real projects, the manifest file is stored inside a remote Git repository that acts as the workspace manifest repository. In that case, initialize the workspace directly from the remote repository:

```
west init -m https://github.com/example/my-manifest-repo
```

How it works...

The location of `west.yml` defines how the workspace is organized. Three main topologies are commonly discussed.

- **T1 – Zephyr-centric topology**
  In this topology, the manifest lives in the Zephyr repository. Zephyr's `west.yml` defines all other repositories. This approach is generally not recommended for application development because:

  It fetches many unused modules

  The application is not part of the manifest

  The workspace is controlled entirely by Zephyr

- **T2 – Application-centric topology (Recommended)**
  In this topology, the application repository contains the main `west.yml` file. The application becomes the central coordination point of the workspace. Zephyr and other dependencies are listed as projects in the application's manifest. This approach is recommended because:

  Only required dependencies are fetched

  The application is version-controlled together with its dependencies

  The workspace structure is explicit and intentional

  Typical layout:

```
west-workspace/
├── .west/
├── my_app/
│   ├── src/
│   ├── prj.conf
│   └── west.yml
└── deps/
    ├── zephyr/
    └── modules/
```

This is the most common and scalable approach for product development.

- **T3 – Manifest-only repository**
  In this topology, the manifest lives in a dedicated repository. Applications, Zephyr, and modules are all siblings. This approach is useful when managing multiple applications that share the same platform, but it adds complexity and is less common for single-product projects.

A Reference Manifest Template

The following example shows a typical application-centric (T2) manifest. It can be used as a starting point for most Zephyr-based projects.

```
manifest:
  remotes:
    - name: zephyrproject
      url-base: https://github.com/zephyrproject-rtos

  projects:
    - name: zephyr
      remote: zephyrproject
      repo-path: zephyr
      revision: v4.4.1
      import:
        path-prefix: deps
        name-allowlist:
          - cmsis
          - hal_nxp
          - hal_stm32
          - percepio

    - name: custom-module-myfeatures
      url: https://github.com/Ac6Embedded/myfeatures
      revision: v1.2.3
      path: deps/modules/lib/myfeatures

  self:
    path: my_app
```

This manifest places the application at the workspace root, fetches Zephyr and selected dependencies, and keeps all third-party code under a dedicated `deps/` directory.

Common Mistakes to Avoid

Some common pitfalls when writing manifests include:

- Fetching all modules instead of using allowlists
- Hardcoding full URLs everywhere instead of using remotes
- Mixing application code into the Zephyr repository
- Treating the manifest as a script rather than a declaration

Keeping the manifest minimal and intentional is key to long-term maintainability.

There's more...

The West manifest supports additional features such as group filtering, conditional imports, and custom command registration. These are useful in advanced scenarios but are not required for most projects.

As projects grow, the manifest becomes an important architectural artifact. Treat it as part of the system design, not just a setup file.

See also

- Zephyr Project example application repository: `https://github.com/zephyrproject-rtos/example-application`
- West workspace concepts and organization: `https://docs.zephyrproject.org/latest/develop/west/workspaces.html`
- West manifest documentation: `https://docs.zephyrproject.org/latest/develop/west/manifest.html`
- Blog: *A practical guide to Zephyr West*: `https://interrupt.memfault.com/blog/practical_zephyr_west`
- West: Explained in Simple Words: `https://www.youtube.com/watch?v=LVfUP_S5u-I`

Get this book's PDF copy, code bundle, and more

Scan the QR code (or go to `packtpub.com/unlock`). Search for this book by name, confirm the edition, and then follow the steps on the page.

UNLOCK NOW

Note: Have your invoice handy. Purchases made directly from the Packt website don't require an invoice.

3

Configuring Zephyr Applications with Kconfig

Zephyr uses Kconfig as its primary configuration system to control which features are enabled, how subsystems behave, and how the final firmware image is built. Instead of hardcoding feature choices into source files, developers can use Kconfig to manage configuration in a structured and scalable way, making it easier to adapt applications across boards and use cases.

In practice, Zephyr applications often need to enable features such as logging, networking, or debugging support while keeping unnecessary code out of the final firmware. Without a clear understanding of how Kconfig works, it can be difficult to track configuration changes, manage dependencies, or explain why certain features are included in a build.

This chapter focuses on practical techniques for working with Kconfig in Zephyr applications. You will learn how to control features using configuration symbols, inspect the generated configuration, create custom symbols, and use Kconfig to influence both source code and build behavior. The recipes also demonstrate how to interact with configuration tools and make configuration changes persistent across builds.

Together, these recipes provide a structured approach to understanding and managing Zephyr application configuration using Kconfig.

In this chapter, we will cover the following recipes:

- Controlling Zephyr features using Kconfig
- Generating the merged Kconfig configuration file
- Adding custom Kconfig symbols to a Zephyr application
- Controlling source compilation with Kconfig

- Modifying configuration using `menuconfig` and `guiconfig`
- Creating persistent Kconfig configuration

Technical requirements

To follow the recipes in this chapter, you will need:

- A supported host machine
- Zephyr SDK or a compatible toolchain
- A supported development board (e.g., STM32, NXP) or a simulation environment
- Visual Studio Code with Workbench for Zephyr extension (optional but recommended)

The code files for this chapter are available in the downloadable code bundle under the Chapter 3 folder. You can download the code by following the instructions in the **Download the example code files** section in the *Preface*.

Controlling Zephyr features using Kconfig

Zephyr applications often need to enable features such as logging, networking, or debugging support while keeping unnecessary code out of the firmware image. Manually editing source files for every build variant would quickly become difficult to maintain.

In this recipe, you will use Kconfig symbols in `prj.conf` to enable and disable Zephyr features, rebuild the application, and observe how configuration choices affect the generated firmware.

Getting ready

- A working Zephyr installation (west workspace set up)
- Any supported board
- The ability to build and run a Zephyr sample
- We will use the Zephyr Hello World sample as the starting point

When you build any Zephyr application, the final configuration will be generated inside the build directory, typically under:

```
<build-dir>/zephyr/.config
<build-dir>/zephyr/include/generated/autoconf.h
```

How to do it...

1. **Create a new application from Hello World**

Open the Create New Application panel in Workbench for Zephyr.

- Select your target board
- Select Sample project: `Hello World`
- In Project Name, enter: `kconfig_hello`
- Finish the wizard

This is functionally equivalent to copying the sample manually:

```
cp -r zephyr/samples/hello_world kconfig_hello
```

But using the Workbench panel ensures the project is properly registered in your workspace.

After creation, your project structure should look like this:

```
kconfig_hello/
├── CMakeLists.txt
├── prj.conf
└── src/main.c
```

2. **Building the project (baseline)**
   Right-click on the `kconfig_hello` application in the Project Explorer and build it.

   This is equivalent to running:

```
west build -b <your_board> kconfig_hello
```

   At this stage, the default Hello World application builds and runs normally. No custom configuration has been added yet.

3. **Enabling Zephyr features in prj.conf**
   Open `prj.conf` and add:

```
CONFIG_LOG=y
CONFIG_LOG_DEFAULT_LEVEL=3
```

   These options:

   - Enable the logging framework
   - Set the default logging level

   Save the file.

4. **Making the application react to Kconfig**

   Modify `src/main.c` to check which features are enabled:

```
#include <zephyr/kernel.h>

int main(void)
{
    printk("Hello World from Zephyr!\n");

#if defined(CONFIG_LOG)
    printk("CONFIG_LOG is enabled (level=%d)\n", CONFIG_LOG_DEFAULT_LEVEL);
#else
    printk("CONFIG_LOG is disabled\n");
#endif

    return 0;
}
```

   This code uses conditional compilation to detect whether the features are enabled.

   Rebuild the project, then flash and run it.

   You should now see a message on the serial output confirming that logging is enabled.

5. **Disabling a feature and observing the effect**

   Now modify `prj.conf`, and comment the options:

```
# CONFIG_LOG=y
# CONFIG_LOG_DEFAULT_LEVEL=3
```

   By removing the request to enable logging, the symbol falls back to its default value. This is equivalent to:

```
CONFIG_LOG=n
```

   Rebuild the project. Flash and run the application again.

   You should now see:

```
CONFIG_LOG is disabled
```

If you inspect the generated file:

```
<build-dir>/zephyr/.config
```

You will find:

```
# CONFIG_LOG is not set
```

This confirms that the logging subsystem was not compiled into the firmware.

How it works...

When you build a Zephyr application, the build system first runs the configuration phase.

During this phase:

- The board default configuration is loaded
- Your `prj.conf` is merged
- Dependencies are resolved
- A final `.config` file is generated

The resulting configuration is stored in:

```
<build-dir>/zephyr/.config
```

For example, when logging is enabled, you will find:

```
CONFIG_LOG=y
CONFIG_LOG_DEFAULT_LEVEL=3
```

The build system also generates:

```
<build-dir>/zephyr/include/generated/autoconf.h
```

In this header, configuration symbols become C macros:

```
#define CONFIG_LOG 1
#define CONFIG_LOG_DEFAULT_LEVEL 3
```

That is why `#if defined(CONFIG_LOG)` works. The macro exists only if the feature is enabled.

Kconfig symbols therefore control what gets compiled into your final firmware image.

There's more...

This is pure compile-time conditional compilation. The code inside the block does not exist at all when the symbol is disabled.

Zephyr also provides a more flexible macro:

```
IS_ENABLED(CONFIG_SYMBOL)
```

This allows you to write regular "if" statements that still get optimized at compile time.

You can write:

```
if (IS_ENABLED(CONFIG_LOG)) {
    printk("Logging is enabled\n");
} else {
    printk("Logging is disabled\n");
}
```

> **Note**
>
> When `CONFIG_LOG` is disabled (n), the `macro CONFIG_LOG` is not defined at all. Writing if (`CONFIG_LOG`) would make the compiler see if (`/* nothing */`), which causes a compile-time error. Using `IS_ENABLED(CONFIG_LOG)` avoids this problem because it always expands to 1 if enabled or 0 if disabled, ensuring safe and clean conditional checks.

See also

- Zephyr Project Documentation, Kconfig: `https://docs.zephyrproject.org/latest/build/kconfig/index.html`
- Zephyr Project Documentation, Setting Kconfig configuration values: `https://docs.zephyrproject.org/latest/build/kconfig/setting.html`

Generating the merged Kconfig configuration file

Zephyr applications often rely on multiple kconfig configuration sources. Board defaults, application settings, and optional fragments are merged during the build process. Without understanding this sequence, it can be difficult to explain why a symbol is enabled or where a final value comes from.

In this recipe, you will inspect the configuration inputs used by Zephyr, verify their precedence, and observe how they generate the final .config and `autoconf.h` files.

Getting ready

You need:

- The `kconfig_hello` application
- A successful build
- Access to the build directory

After building, verify that .config and `autoconf.h` exist.

```
build-dir/zephyr/.config
build-dir/zephyr/include/generated/autoconf.h
```

How to do it...

1. **Identify the configuration inputs**
   Zephyr does not generate .config from `prj.conf` alone.

   It merges:

   - `<BOARD>_defconfig`
   - `prj.conf`
   - Optional extra fragment files (if defined using `EXTRA_CONF_FILE`)

2. **Locate the board default configuration**
   Find the board's default configuration:

   ```
   boards/<arch>/<board>/<board>_defconfig
   ```

   This file defines the baseline configuration required for the board.

   It is always processed first.

3. **Verify precedence with a controlled change**
   Choose a symbol that is enabled in `<BOARD>_defconfig`.

   Now explicitly set a different value in `prj.conf`.

   Rebuild the project.

4. **Inspect the final result**

   Open:

```
<build-dir>/zephyr/.config
```

   You will see that the value from `prj.conf` overrides the board default.

   This confirms the precedence rule: application configuration overrides board defaults.

How it works...

The configuration phase follows a strict sequence:

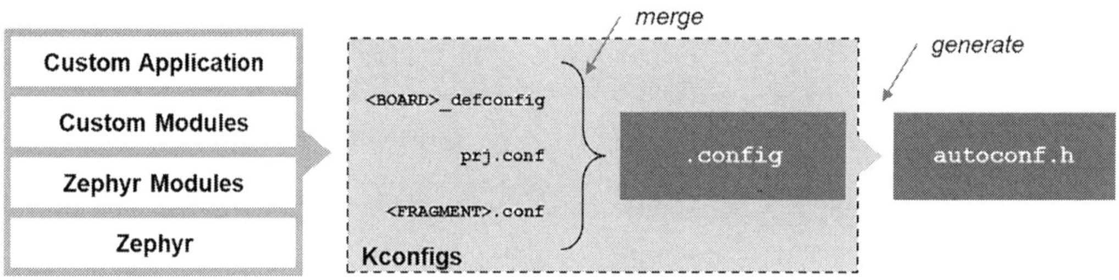

Figure 3.1 – Kconfig merge flow in Zephyr

The process is:

1. **Load <BOARD>_defconfig**
2. **Apply prj.conf**
3. **Apply extra fragments**
4. **Resolve dependencies and defaults**
5. **Write .config**
6. **Generate autoconf.h**

Precedence rule:

- Later inputs override earlier ones
- Dependency rules may still force a final value

The final decision is always visible in .config. The `autoconf.h` file converts enabled symbols into C macros used during compilation. Later inputs usually override earlier ones, but dependency rules may still force a different final value.

There's more...

You can observe the exact merge sequence directly in the build log on the console.

During the configuration phase, Zephyr prints messages similar to:

```
Parsing .../zephyr/Kconfig
Loaded configuration '.../<board>_defconfig'
Merged configuration '.../prj.conf'
Merged configuration '.../boards/<board>.conf'
Merged configuration '.../my_extra.conf'
Configuration saved to '.../build/zephyr/.config'
Kconfig header saved to '.../build/zephyr/include/generated/autoconf.h'
```

The `boards/<board>.conf` file does not need to be explicitly declared.

If it exists inside your application, Zephyr detects and merges it automatically for that specific board.

See also

- Practical Zephyr - Kconfig: `https://interrupt.memfault.com/blog/practical_zephyr_kconfig`

Adding custom Kconfig symbols to a Zephyr application

Built-in Zephyr configuration symbols cover many common features, but applications often need their own feature toggles, selectable options, or custom settings. Without application-level symbols, these choices usually end up hardcoded in source files.

In this recipe, you will create a custom Kconfig file, define a new symbol named `CONFIG_TOTO`, and integrate it into the Zephyr configuration system.

Getting ready

Create a new Hello World-based application named `kconfig_toto`. The project should be built successfully before adding custom Kconfig symbols.

How to do it...

1. **Create the application Kconfig file**

   At the root of the application (next to `CMakeLists.txt`):

```
kconfig_toto/
├── CMakeLists.txt
├── Kconfig
├── prj.conf
└── src/
    └── main.c
```

   Open Kconfig and add:

```
mainmenu "My ToTo App Configuration"

source "Kconfig.zephyr"

menu "ToTo Application Options"

config TOTO
    bool "Enable ToTo feature"
    default n
    help
        Enables the toto dummy feature in this application.

endmenu
```

2. **Enable the symbol**

   In `prj.conf`:

```
CONFIG_TOTO=y
```

3. **Rebuild the application and verify the result**

   Perform a pristine rebuild so the new Kconfig symbols are detected correctly.

   Command line:

```
west build -p always -b <your_board>
```

Or, in Workbench for Zephyr, right-click on the application, then select Rebuild Pristine.

Now .config contains:

```
CONFIG_TOTO=y
```

The custom symbol is fully integrated into Zephyr's configuration.

How it works...

When present in the application root directory, the Kconfig file becomes the application entry point for custom symbols.

Therefore, source "Kconfig.zephyr" is mandatory to include Zephyr's configuration tree.

The "config TOTO" declaration adds a new symbol to that tree.

During configuration, `prj.conf` assigns a value to `CONFIG_TOTO` just like any other Zephyr symbol.

Once generated, the symbol can be used in C source files, `CMakeLists.txt`, and interactive configuration tools.

There's more...

You can define multiple symbols in the same file and organize them using menu and submenu blocks. Keep application-specific symbols inside the project directory. Avoid modifying Zephyr's internal Kconfig files for application features, as this makes upgrades harder to maintain.

See also

- Add Custom KConfig Symbols in Zephyr: `https://www.zephyrproject.org/add-custom-kconfig-symbols-in-zephyr/`

Controlling source compilation with Kconfig

Kconfig symbols do more than enable code inside C source files. They can also control which source files, include directories, or feature modules are compiled into the application. This is useful when optional features should disappear completely from the build when disabled.

In this recipe, you will use `CONFIG_TOTO` inside `CMakeLists.txt` to conditionally compile source files and include directories.

Getting ready

Use the existing `kconfig_toto` application created in the previous recipe.

The custom symbol `CONFIG_TOTO` should already be available.

How to do it...

1. **Enable the symbol**

   In `prj.conf`:

```
CONFIG_TOTO=y
```

   Rebuild once to apply the configuration.

2. **Create a feature directory with its own CMakeLists.txt**

   Create this structure:

```
kconfig_toto/
├── CMakeLists.txt
├── prj.conf
├── src/
│   └── main.c
└── toto/
    ├── CMakeLists.txt
    ├── toto.c
    └── include/
        └── toto.h
```

`toto/include/toto.h`:

```
void toto_init(void);
```

`toto/toto.c`:

```
#include <zephyr/kernel.h>
#include "toto.h"

void toto_init(void)
{
    printk("Hello from toto! \n");
}
```

3. **Conditionally compile the feature**

   In `toto/CMakeLists.txt`:

```
zephyr_sources_ifdef(CONFIG_TOTO toto.c)
zephyr_include_directories_ifdef(CONFIG_TOTO include)
```

   In the top-level `CMakeLists.txt`, add the directory to the build:

```
add_subdirectory(toto)
```

4. **Use the feature**

   In `src/main.c`:

```
#if defined(CONFIG_TOTO)
#include "toto.h"
#endif

int main(void)
{
#if defined(CONFIG_TOTO)
    toto_init();
#endif
    return 0;
}
```

5. **Rebuild and run**

   If `CONFIG_TOTO=y`, the application prints:

```
Hello from toto!
```

   If disabled, the `toto.c` file is not compiled.

How it works...

During the configuration phase, Zephyr exports `CONFIG_TOTO` to CMake.

The function evaluates the symbol at configure time:

```
zephyr_sources_ifdef(CONFIG_TOTO toto.c)
```

If enabled, the file becomes part of the application target.

If disabled, it is excluded before compilation begins.

The decision is made during build configuration, not at runtime.

There's more...

If a feature contains many files, exclude the entire directory instead:

```
add_subdirectory_ifdef(CONFIG_TOTO toto)
```

In this case, the directory is not processed at all when disabled.

Use:

- `zephyr_sources_ifdef()` for file-level control
- `add_subdirectory_ifdef()` to completely exclude a feature module

Both approaches rely on the same configuration symbol.

See also

- `https://docs.zephyrproject.org/latest/develop/api/design_guidelines.html#conditional-data-and-apis`

Modifying configuration using menuconfig and guiconfig

Many configuration symbols are difficult to discover by editing `prj.conf` manually. Some options depend on others, some are hidden until prerequisites are enabled, and some require help text to understand their purpose.

In this recipe, you will use `menuconfig` and `guiconfig` to browse available symbols, search for options, change values interactively, and save the resulting configuration.

Getting ready

Use the existing `kconfig_toto` application. The project should build successfully before opening the interactive tools.

How to do it...

1. **Open menuconfig**
   From Workbench for Zephyr:

   Right-click the application → Configure → menuconfig

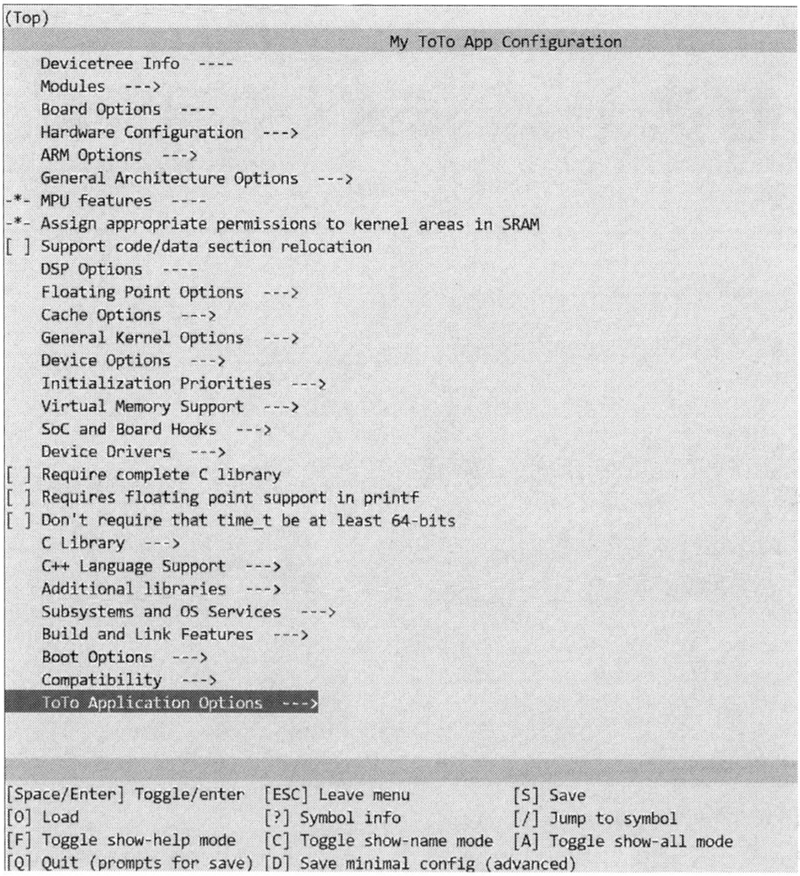

Figure 3.2 – My ToTo App Configuration

Alternatively, from the command line:

```
west build -t menuconfig
```

Basic controls:

- Arrow keys → navigate
- *Enter* → open menu
- *Space / Y / N* → toggle boolean
- / → search
- *?* → show help
- *Q* → save and quit

2. **Search for the custom symbol**

   Press / and search for:

```
TOTO
```

   Enable or disable the symbol.

   When exiting, press *Y* to save.

3. **Using guiconfig**

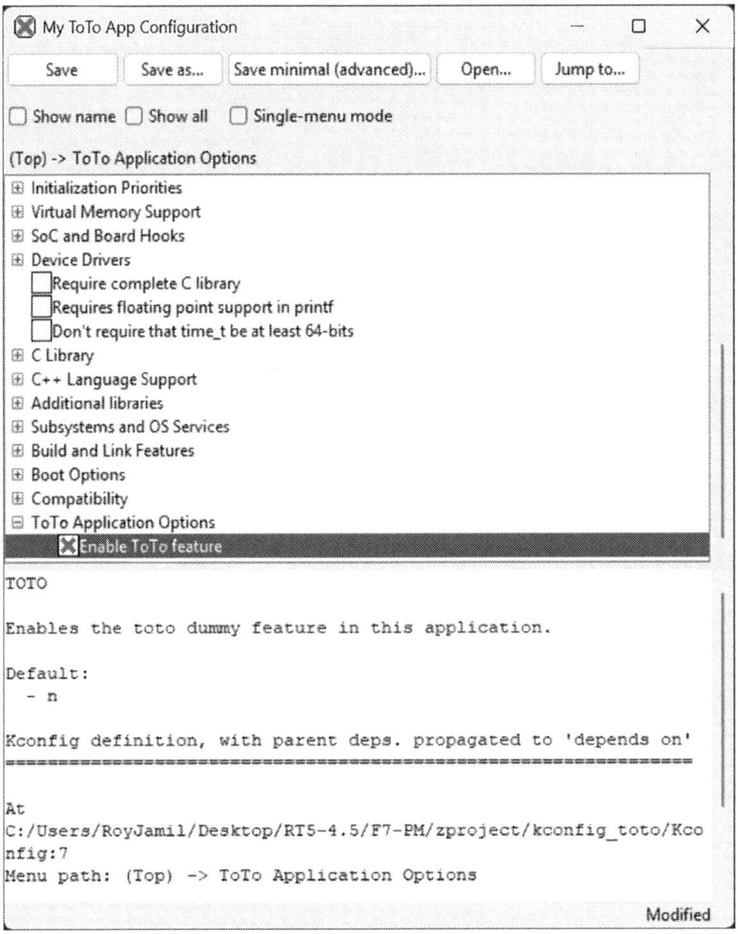

Figure 3.3 – My ToTo App Configuration

The same configuration can also be done using `guiconfig`, the only difference is that you can use the mouse when modifying the options.

How it works...

The configuration tools, `menuconfig` and `guiconfig`, load the complete Kconfig tree for the active build.

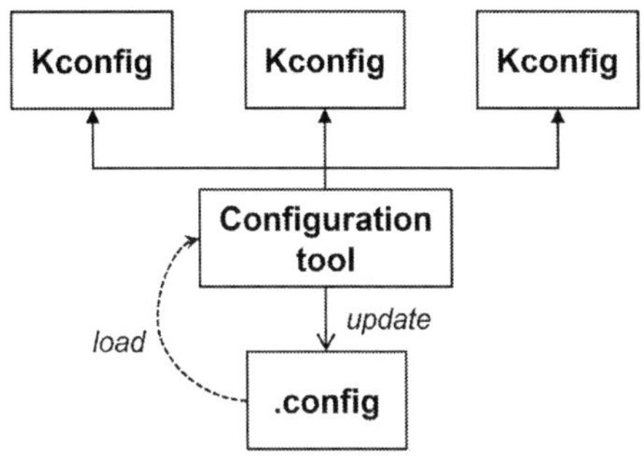

Figure 3.4 – Configuration tool loading Kconfig files and updating .config

The process is:

- Load all Kconfig files
- Evaluate symbol dependencies
- Prevent invalid selections
- Write the result to `.config`

If a symbol has unsatisfied dependencies, it cannot be enabled.

Interactive changes affect only `<build-dir>/zephyr/.config`, to make changes permanent, they must be copied into `prj.conf`.

There's more...

You can save a minimal configuration from:

- `menuconfig` by pressing D
- `guiconfig` by clicking the corresponding button

This generates a "defconfig" which is a configuration file that contains only the symbols whose values differ from their default values.

For example, CONFIG_MULTITHREADING is enabled by default, so it will not appear in the minimal configuration file. If you enable CONFIG_TOTO, it will appear in the minimal configuration file because its value differs from the default.

Use this when:

- Creating a configuration fragment
- Identifying which symbols were modified
- Cleaning up experimental changes
- Creating a custom board defconfig

See also

- Zephyr Project Documentation, Interactive Kconfig interfaces: https://docs.zephyrproject.org/latest/build/kconfig/menuconfig.html
- Configure Zephyr: Kconfigs and Devicetree in Simple Words: https://www.youtube.com/watch?v=XimbF9M-kIU

Creating persistent Kconfig configuration

Interactive tools such as menuconfig modify only the generated .config file inside the build directory. Those changes are temporary and disappear after a pristine rebuild or when the build directory is removed.

In this recipe, you will store configuration changes permanently using prj.conf, board-specific configuration files, and optional configuration fragments.

Getting ready

Use the existing kconfig_toto application. You should already have tested configuration changes with menuconfig or guiconfig.

How to do it...

1. **Store settings in prj.conf**

   Open prj.conf and add:

   ```
   CONFIG_TOTO=y
   ```

   Rebuild the project.

   The setting now persists across clean builds.

   prj.conf is the primary and recommended location for application configuration.

2. **Create a board-specific configuration file**
   For settings that apply only to one board, create:

   ```
   boards/<board>.conf
   ```

   Example:

   ```
   boards/nucleo_f722ze.conf
   ```

   The file is detected automatically when building for the matching board.

   Good practice:

   - Use it only for board-specific configuration
   - Keep application logic inside `prj.conf`

   Do not mix board policy with application features.

3. **Create an extra configuration fragment**
   Create a fragment file, for example:

   ```
   toto.conf
   ```

   Content:

   ```
   CONFIG_TOTO=y
   ```

4. **Register the fragment**
   In `CMakeLists.txt` add:

   ```
   set(EXTRA_CONF_FILE ${CMAKE_CURRENT_SOURCE_DIR}/toto.conf)
   ```

   > **Note**
   >
   > `set(EXTRA_CONF_FILE ...)` must be placed before `find_package(Zephyr ...)`, since Zephyr reads it during configuration. If you place it after, the fragment is silently ignored.

   Fragments are useful when:

   - Creating build variants

- Grouping optional feature sets
- Keeping `prj.conf` minimal

Perform a pristine rebuild or use **Clean Rebuild** in Workbench for Zephyr.

How it works...

During configuration, Zephyr merges several inputs in order:

- Board `<board>_defconfig`
- `prj.conf`
- `boards/<board>.conf` (if present)
- Extra fragments (if present)

The result generates a new .config. Because `prj.conf` and fragment files belong to the application source tree, their settings persist across clean builds.

There's more...

How to know which symbols to store:

You typically determine the required symbols in one of three ways:

- From official documentation
- By inspecting a similar Zephyr sample
- By experimenting with `menuconfig`

A common workflow when experimenting with `menuconfig` is:

1. **Save a minimal configuration of the original state**
2. **Enable the required features interactively**
3. **Save a minimal configuration again**
4. **Compare both minimal configuration files**
5. **Store only the symbols that differ from their default values**

This approach avoids copying unnecessary options and keeps your persistent configuration clean and maintainable.

See also

Configure Zephyr: Kconfigs and Devicetree in Simple Words: `https://www.youtube.com/watch?v=XimbF9M-kIU`

4

Using Devicetree to Describe Hardware

Zephyr uses devicetree to describe hardware at build time. Instead of hardcoding peripheral addresses, pin assignments, or interrupt numbers in source code, hardware configuration is defined separately and transformed into compile-time macros. This separation allows applications to remain portable while drivers adapt to different boards and **SoCs**.

In this chapter, we will cover the following recipes:

- Inspecting Devicetree in Zephyr
- Creating Devicetree overlays
- Customizing Devicetree nodes and properties
- Using Devicetree in application code
- Writing portable code with aliases

Technical requirements

To follow the recipes in this chapter, you will need:

- A supported host machine
- Zephyr SDK or a compatible toolchain
- A supported development board (e.g., STM32, NXP) or a simulation environment
- Visual Studio Code with Workbench for Zephyr extension (optional but recommended)
- Preferably, the OpenSmart Rich Shield, which provides multiple peripherals used throughout the book, including a potentiometer

- Otherwise, a potentiometer connected to the Arduino A0 pin of the development board
- Optionally, an external LED (with a current-limiting resistor) connected to a free GPIO pin, the onboard LEDs are sufficient if available.

The code files for this chapter are available in the downloadable code bundle under the Chapter 4 folder. You can download the code by following the instructions in the **Download the example code files** section in the *Preface*.

Inspecting Devicetree in Zephyr

Embedded applications often need to run on different boards and microcontrollers. Without a separate hardware description system, peripheral addresses, interrupts, and pin assignments would need to be hardcoded directly in source code, making applications difficult to port and maintain.

Zephyr uses devicetree to describe hardware separately from application logic. This allows the same application code to run across multiple boards while hardware-specific details are selected at build time.

In this recipe, you will inspect the devicetree files used by Zephyr, identify the generated merged devicetree, and observe how hardware information becomes available to drivers and applications.

Getting ready

Use any existing Zephyr application, such as `hello_world`.

Build or rebuild your project either via Workbench for Zephyr by doing "Pristine build/ Rebuild" or from the command line:

```
west build -b <your_board> -p
```

During the configuration phase, the build output shows which board Devicetree file is used. For example:

```
-- Found BOARD.dts: .../boards/st/nucleo_h563zi/nucleo_h563zi.dts
-- Generated zephyr.dts: .../build/zephyr/zephyr.dts
```

These are the main files inspected in this recipe.

How to do it...

1. **Inspect the generated devicetree**

   Open:

```
build/zephyr/zephyr.dts
```

   This file contains the final merged devicetree used during the build.

2. **Locate a peripheral node**

   Search for a UART node.

   You may find something similar to:

```
uart0: serial@40004400 {
    compatible = "st,stm32-uart";
    reg = <0x40004400 0x400>;
    interrupts = <53 0>;
    status = "okay";
};
```

   This is a definition of node, where we specify the source code of the driver that is going to manage it via compatible property. And other properties to configure it like the memory mapped base address and the interrupt number and priority.

3. **Inspect aliases**

   Search for:

```
aliases {
```

   You may find:

```
aliases {
    led0 = &green_led;
    sw0 = &user_button;
};
```

   Aliases provide stable application-facing names for hardware nodes.

   Applications can use led0 or sw0 instead of board-specific node labels.

How it works...

Without devicetree, hardware might be defined directly in code.

VendorA:

```
#define UART_BASE 0x40021000
#define UART_IRQ  37
```

This works for one board.

Now imagine switching to a board from another vendor.

VendorB:

```
#define UART_BASE 0x50003000
#define UART_IRQ  12
```

Every time you change boards, you must modify source code.

This does not scale and breaks portability.

With devicetree, each vendor describes hardware in its own `.dts` file.

VendorA:

```
uart0: uart@40021000 {
    compatible = "vendora,uart";
    reg = <0x40021000 0x400>;
    status = "okay";
};
```

VendorB:

```
uart0: uart@50003000 {
    compatible = "vendorb,uart";
    reg = <0x50003000 0x400>;
    status = "okay";
};
```

The hardware description differs, but the application code does not change time.

This approach is lightweight and efficient for embedded systems.

There's more...

A Zephyr board usually does not describe all hardware inside a single `.dts` file.

The board devicetree file:

```
boards/<vendor>/<board>/<board>.dts
```

typically includes additional `.dtsi` files describing:

- The SoC
- Pin control configuration
- Shared peripherals
- Common board features

For example:

```
#include <st/h5/stm32h563xx.dtsi>
```

This layered organization allows multiple boards to reuse the same SoC hardware description.

After all files are merged, Zephyr generates:

```
build/zephyr/zephyr.dts
```

This file represents the final hardware description used during the build, it is useful for debugging.

See also

- Zephyr Project Documentation – Devicetree guide: `https://docs.zephyrproject.org/latest/build/dts/index.html`
- Configure Zephyr: Kconfig and Devicetree Explained: `https://www.youtube.com/watch?v=XimbF9M-kIU`

Creating Devicetree overlays

Board devicetree files define the default hardware configuration for a board. However, applications often need to enable peripherals, modify properties, or add hardware descriptions without changing the original board files directly.

Devicetree overlays solve this problem by allowing applications to extend or override the hardware configuration at build time.

In this recipe, you will create devicetree overlays, observe how Zephyr detects them during the build, and inspect the generated merged devicetree output.

Getting ready

Rebuild the project with a pristine build.

During the configuration phase, observe the devicetree processing messages in the build output. These messages show the exact files being used:

```
-- Found BOARD.dts: …/boards/st/nucleo_h563zi/nucleo_h563zi.dts
-- Generated zephyr.dts: …/build/primary/zephyr/zephyr.dts
-- Generated devicetree_generated.h: …/zephyr/devicetree_generated.h
```

How to do it...

1. **Create a board overlay**
   Inside your application, create a subfolder named boards, and add a file named: `<board>.overlay`, where `<board>` matches the exact board name (for example, `nucleo_h563zi.overlay`).

   This file should contain hardware customizations specific to that board.

   No additional registration is required for now.

2. **Rebuild the application**
   Perform a pristine rebuild.

   The build system automatically detects this overlay when building for the matching board.

   Even an empty file is sufficient to test detection.

   You should see a message similar to:

   ```
   -- Found devicetree overlay: …/boards/nucleo_h563zi.overlay
   ```

3. **Create an application overlay (optional)**
   Alternatively, you can create at the root of your application a file named: `app.overlay`.

   When both `app.overlay` and `boards/<board>.overlay` exist, only the board-specific overlay is used.

4. **Create an extra overlay (optional)**

   Most applications only need a board-specific overlay such as:

```
boards/<board>.overlay
```

   This is usually sufficient for enabling peripherals or adjusting common hardware settings.

   For larger applications with more complex hardware customization, additional overlay files can also be provided using:

```
EXTRA_DTC_OVERLAY_FILE=path/to/myfile.overlay
```

How it works...

Devicetree overlays do not replace the board devicetree source file, they extend or override it.

Because overlays are applied last, they can:

- Change property values
- Modify node status
- Add new nodes

The merge happens entirely at build time. No runtime modification is possible.

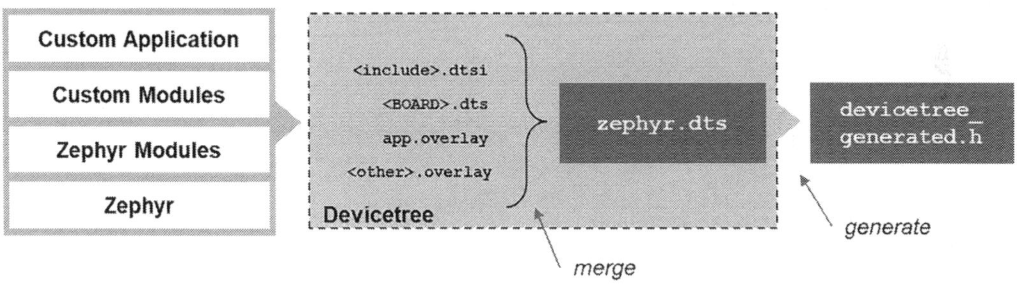

Figure 4.1 – Generation flow of zephyr.dts and devicetree_generated.h

The process is:

1. Devicetree sources are merged into a single tree.
2. The merged result becomes `zephyr.dts`.
3. The build system generates `devicetree_generated.h`.

The file `zephyr.dts` represents the final, fully merged hardware description used for the build. It is useful for inspection and debugging.

The file `devicetree_generated.h` contains C macros created from the merged devicetree. These macros are included automatically during compilation.

They allow your code to:

- Refer to a hardware node by name
- Read values defined in the devicetree
- Check whether a device is enabled

Your application and drivers do not read `.dts` files directly.

They use these generated macros instead.

There's more...

The most useful file to inspect after applying an overlay is:

```
build/zephyr/zephyr.dts
```

This file shows the final merged devicetree after the board files, SoC files, and overlays have been combined. If an overlay does not seem to work, first check this file to confirm that the expected node or property is present.

For example, if you enabled a peripheral in an overlay, verify that the final node contains:

```
status = "okay";
```

You can also inspect:

```
build/zephyr/include/generated/devicetree_generated.h
```

This file contains the generated C macros used by drivers and application code.

See also

- Zephyr Project Documentation, Devicetree guide: `https://docs.zephyrproject.org/latest/build/dts/index.html`
- Zephyr Project Documentation, Devicetree overlays: `https://docs.zephyrproject.org/latest/build/dts/howtos.html#set-devicetree-overlays`
- Interrupt Blog, Practical Zephyr Devicetree Overview: `https://interrupt.memfault.com/blog/practical_zephyr_dt`

- Mastering Zephyr Driver Development: `https://www.youtube.com/watch?v=o-f2qCd2AXo`
- Configure Zephyr: Kconfigs and Devicetree in Simple Words: `https://www.youtube.com/watch?v=XimbF9M-kIU`

Customizing Devicetree nodes and properties

Board devicetree files already define most peripherals and hardware settings. However, applications often need to enable a disabled peripheral, modify an existing property, or add application-specific hardware configuration without rewriting the entire node.

Devicetree overlays solve this problem by allowing applications to update only the required parts of an existing hardware description.

In this recipe, you will modify an existing devicetree node, update its properties, and verify the changes in the generated devicetree output.

Getting ready

You must already have the board overlay created in the previous recipe.

Hardware setup:

- If you use a shield, the ADC input is already connected.
- Otherwise, connect a potentiometer to the Arduino header A0 pin (and connect GND and VCC accordingly).

On STM32H5 Nucleo, Arduino A0 is connected to ADC1 channel 3 as shown in this screenshot from the STM32H5 Nucleo user manual.

Zio pin	Pin name	Signal name	STM32 pin	STM32 function
1	A0	ADC	PA6	ADC12_INP3
3	A1	ADC	PC0	ADC12_INP10
5	A2	ADC	PC3	ADC12_INP13
7	A3	ADC	PB1	ADC12_INP5
9	A4	ADC	PC2/ PB9	ADC12_INP12/ I2C1_SDA
11	A5	ADC	PF11/P B8	ADC1_INP2/ I2C1_SCL

Figure 4.2 – Arduino connector pins on the STM32H5 Nucleo board

On FRDM-MCXN236, the Arduino A0 input used in this recipe is connected to PIO4_6 on the MCXN236 and routed to LPADC0 channel 3 input A. This is shown in the FRDM-MCXN236 user manual.

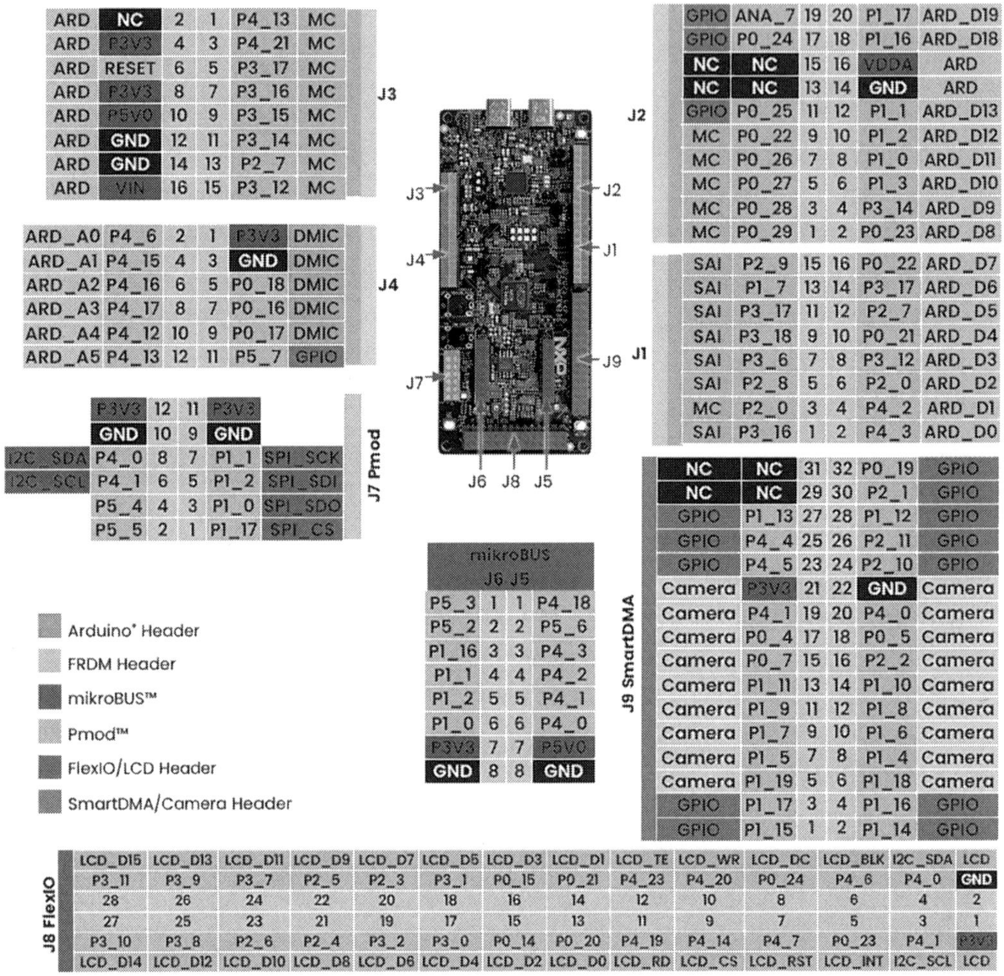

Figure 4.3 – Arduino connector pins on the FRDM-MCXN236 board

How to do it...

1. **Locate the existing peripheral node**

   The ADC peripheral is already defined in the board devicetree file, typically inside:

   The board `.dts` file, or one of the included `.dtsi` files

   You can verify this by opening the board `.dts` file and searching for adc.

For example, on an ST board, you may find something similar to:

```
&adc1 {
    pinctrl-0 = <&adc1_inp3_pa6 &adc1_inp15_pa3>;
    st,adc-prescaler = <4>;
    status = "okay";
};
```

This means the ADC hardware instance not only exists but is already enabled, with a default pin configuration.

On NXP FRDM MCXN236, `lpadc0`:

```
lpadc0: lpadc@10d000 {
    compatible = "nxp,lpc-lpadc";
    status = "okay";
    pinctrl-0 = <&pinmux_lpadc0>;
};
```

2. **Create or open the board overlay**
   Open:

```
boards/<board>.overlay
```

You do not redefine nodes from the root.

Overlay content does not require wrapping inside `/{}` when using `&label`.

Add:

```
&adc1 {
    pinctrl-0 = <&adc1_inp3_pa6>;
};
```

Here:

- `&adc1` references the existing node
- `pinctrl-0 = <&adc1_inp3_pa6>` overrides the board's default `pinctrl` list, keeping only the pin connected to Arduino A0.

On FRDM MCXN236, you need to reconfigure the pinmux:

```
#include <zephyr/dt-bindings/adc/adc.h>
#include <zephyr/dt-bindings/adc/mcux-lpadc.h>

&pinmux_lpadc0 {
    group0 {
        pinmux = <ADC0_A3_PIO4_6>;
    };
};
```

No need to change the `pinctrl-0` property as it is already referring to `pinmux_lpadc0` by default.

You are updating the existing node defined in the board `.dts` file.

Add a custom property (optional)Some drivers require additional configuration.

For example, ST ADC may need vendor-specific properties such as:

```
&adc1 {
    status = "okay";
    st,adc-prescaler = <4>;
};
```

Vendor-specific properties start with the vendor prefix (for example, st,).

If a property is not valid, the build fails during configuration.

3. **Add a channel subnode**
   Beyond properties on the ADC node itself, the driver expects one child node per channel the application reads from. The child node describes how that specific input should be sampled.

   Inside `&adc1 { ... }`, add the following subnode:

```
&adc1 {
    channel@3 {
        reg = <3>;
        zephyr,gain = "ADC_GAIN_1";
        zephyr,reference = "ADC_REF_INTERNAL";
        zephyr,acquisition-time = <ADC_ACQ_TIME_DEFAULT>;
        zephyr,resolution = <12>;
```

```
    };
};
```

Here is what each piece does:

`channel@3` is the child node. The `@3` is its unit address, the channel number, and it must match the reg property.

`zephyr,resolution` sets the conversion resolution, in bits.

`zephyr,gain`, `zephyr,reference`, and `zephyr,acquisition-time` come from the generic Zephyr ADC channel binding. You can see how they are interpreted on this SoC in the STM32 ADC driver at `adc_stm32.c`.

For the FRDM MCXN236:

```
&lpadc0 {
    #address-cells = <1>;
    #size-cells = <0>;

    channel@3 {
        reg = <3>;
        zephyr,gain = "ADC_GAIN_1";
        zephyr,reference = "ADC_REF_EXTERNAL1";
        zephyr,acquisition-time = <ADC_ACQ_TIME_DEFAULT>;
        zephyr,vref-mv = <1800>;
        zephyr,resolution = <12>;
        zephyr,input-positive = <MCUX_LPADC_CH3A>;
    };
};
```

It is worth noting that compared to the STM32 overlay, `zephyr,vref-mv = <1800>` is defined inside the channel instead of the ADC node.

If you need to read several channels, declare one `channel@<n>` subnode per channel, each with its own reg value matching the channel index.

4. **Rebuild and verify the generated file**
   Perform a pristine rebuild. After the build, open:

```
build/zephyr/zephyr.dts
```

You should see that adc1 now has `status = "okay"`.

How it works...

A devicetree node follows a strict structure:

Figure 4.4 – Structure of a devicetree node and its properties

`adc1` is the node label. It allows references using `&adc1`.

`adc@42028000` is the node name and unit address.

Properties:

- Always end with ;
- Use = for assignment
- Use < > for numeric values
- Use quotes for strings

If syntax is incorrect, the build fails during configuration.

There's more...

A property name is not "free text".

Most properties are checked against a binding file. This file defines what is allowed, what is required, and what values are accepted.

For the STM32 ADC driver, the binding is matched using:

```
compatible: "st,stm32-adc"
```

Bindings can also mark properties as required. For example, the STM32 ADC binding marks the prescaler as required:

```
st,adc-prescaler:
  type: int
  required: true
```

This means:

If you forget `st,adc-prescaler`, the build will fail.

If you set an invalid type (for example "four" as a string), the build will fail:

```
&adc1 {
    st,adc-prescaler = "four"; /* will fail at build time */
};
```

This causes a build error because the binding expects an integer.

Bindings can also provide a default value. In the same binding, `vref-mv` defaults to 3300:

```
vref-mv:
  type: int
  default: 3300
```

So if you do not set `vref-mv` in your overlay, the driver still gets a valid value.

This is why bindings are important: they prevent wrong devicetree configurations from silently compiling.

See also

- Zephyr Project Documentation – Devicetree syntax and structure: `https://docs.zephyrproject.org/latest/build/dts/intro-syntax-structure.html`
- Interrupt Blog – Practical Zephyr Devicetree Overview: `https://interrupt.memfault.com/blog/practical_zephyr_dt`
- Configure Zephyr: Kconfigs and Devicetree in Simple Words: `https://www.youtube.com/watch?v=XimbF9M-kIU`

- NUCLEO-H563ZI User Manual: `https://www.st.com/resource/en/user_manual/um3115-stm32h5-nucleo144-board-mb1404-stmicroelectronics.pdf`
- NXP FRDM-MCXN236 documentation: `https://www.nxp.com/design/design-center/development-boards-and-designs/FRDM-MCXN236`

Using Devicetree in application code

Devicetree in Zephyr does not configure hardware dynamically at runtime. Instead, Zephyr converts the hardware description into compile-time macros that applications can access directly from C code.

Applications often need to read values defined in devicetree such as base addresses, channel numbers, resolutions, or vendor-specific settings, without hardcoding them in source code.

In this recipe, you will access devicetree information from application code, navigate between nodes, and read properties as compile-time constants.

Getting ready...

Starting from a basic Hello World project, you will access the ADC node using its label and read one of its properties directly from the devicetree.

Start from a clean Hello World, create a new project from the `hello_world` sample, you can rename it to `my_adc`

Use the same overlay defined in the previous recipe, copy the `boards/<my_board>.overlay`

Rebuild before proceeding.

How to do it...

1. **Include required headers**
   Open `src/main.c` and include:

```
#include <zephyr/kernel.h>
```

2. **Retrieve the ADC node identifier**
   Add the following definition before `main()`, for STM32 Nucleo H5:

```
#define MY_ADC_NODE DT_NODELABEL(adc1)
```

For FRDM MCXN236:

```
#define MY_ADC_NODE DT_NODELABEL(lpadc0)
```

`DT_NODELABEL()` retrieves the node identifier corresponding to the label defined in the devicetree.

3. **Read a simple property**
   For STM32 Nucleo H5, `vref-mv` is defined on the ADC node:

```
#define ADC_VREF_MV DT_PROP(MY_ADC_NODE, vref_mv)
```

For FRDM MCXN236, the same property is defined on the channel subnode:

```
#define ADC_VREF_MV DT_PROP(DT_CHILD(MY_ADC_NODE, channel_3),
zephyr_vref_mv)
```

`DT_PROP()` reads a property from a node. The property name is written with _ instead of -, following the same normalization rule as node names.

4. **Resolve the channel node**
   The ADC channel is a subnode of the ADC. Add:

```
#define MY_ADC_CH_NODE DT_CHILD(MY_ADC_NODE, channel_3)
```

`DT_CHILD()` retrieves the `channel@3` subnode. The @ is normalised to _, which is why the C code writes `channel_3`.

5. **Read a property with a fallback**
   Some properties may not be present on every board. Use `DT_PROP_OR()` to provide a default value:

```
#define ADC_RESOLUTION DT_PROP_OR(MY_ADC_CH_NODE, zephyr_resolution, 12)
```

If zephyr,resolution is defined in the overlay, its value is used. Otherwise, the fallback value 12 is used. This is resolved entirely at compile time.

6. **Check whether the node is enabled**
   You can verify if you set `status = "okay"` to enable the ADC from C using `DT_NODE_HAS_STATUS()`:

```
#define ADC_ENABLED DT_NODE_HAS_STATUS(MY_ADC_NODE, okay)
```

This expands to 1 if the node is present and enabled in the merged devicetree, and 0 otherwise.

7. **Print the values**

   Replace `main()` with:

```
int main(void)
{
    printk("ADC vref:        %d mV\n",   ADC_VREF_MV);
    printk("ADC resolution:  %d bits\n", ADC_RESOLUTION);
    printk("ADC enabled?     %s\n",      ADC_ENABLED ? "yes" : "no");

    return 0;
}
```

8. **Rebuild and flash the application**

   You should see output similar to:

```
ADC vref:        3300 mV
ADC resolution:  12 bits
ADC enabled?     yes
```

   None of these values appears as a literal in your C code. All of them come from devicetree, resolved by the preprocessor before the compiler runs.

How it works...

During the build process, Zephyr merges all devicetree sources and overlays into a single `zephyr.dts` file.

From this merged tree, Zephyr generates the `devicetree_generated.h` header.

This file contains C macros representing every node and property in the tree.

For example, the node:

```
/soc/adc@42028000
```

becomes a macro identifier:

```
DT_N_S_soc_S_adc_42028000
```

The naming rules are:

- Each level in the path is separated using `_S_`
- Non-alphanumeric characters become `_`

- @ becomes _
- \- becomes _

Properties are appended using `_P_`.

For example:

```
vref-mv
```

becomes:

```
_P_vref_mv
```

So the full macro representing the property becomes:

```
DT_N_S_soc_S_adc_42028000_P_vref_mv
```

This is exactly what the macro:

```
DT_PROP(MY_ADC_NODE, vref_mv)
```

expands to internally.

`DT_NODELABEL(adc1)` resolves to the internal node identifier macro.

`DEVICE_DT_GET()` retrieves the struct device instance created by the driver model for that node.

All of this happens at compile time. No devicetree parsing occurs at runtime.

There's more...

The `adc_dt` sample typically expects the application to specify which ADC and channel to read using `/zephyr,user`, for example:

```
/ {
    zephyr,user {
        io-channels = <&adc1 3>;
    };
};
```

The `/zephyr,user` node is meant for application-specific settings. It is a clean place to describe "what this app wants to use" (such as an ADC channel) without changing the board's default hardware description.

You will use this pattern in the next recipe when accessing the ADC from application code.

Rebuild and flash the application.

The `adc_dt` sample will start printing periodic readings. The output looks like this:

```
ADC reading[0]:
- <adc_device_name>, channel 3: 1234 = 995 mV
ADC reading[1]:
- <adc_device_name>, channel 3: 1308 = 1055 mV
ADC reading[2]:
- <adc_device_name>, channel 3: 1180 = 952 mV
```

Turning the potentiometer should change the reported values.

See also

- Zephyr Project Documentation – Devicetree API: `https://docs.zephyrproject.org/latest/build/dts/api/api.html`
- Interrupt Blog – Practical Zephyr Devicetree: `https://interrupt.memfault.com/blog/practical_zephyr_dt`
- Digikey – Devicetree Tutorial: `https://www.digikey.com/en/maker/tutorials/2025/introduction-to-zephyr-part-4-devicetree-tutorial`
- Zephyr Tech Talk – Devicetree internals: `https://www.youtube.com/watch?v=w8GgP3h0M8M`

Writing portable code with aliases

Hardcoding devicetree node labels inside application code reduces portability. Different boards and vendors often use different names for similar peripherals, which forces applications to change when moving between targets. Devicetree aliases solve this problem by providing stable application-facing names that can map to different hardware nodes on different boards.

In this recipe, you will define a devicetree alias and use it in application code to make the application portable across boards.

Getting ready

Reuse the ADC overlay and application created in the previous recipes.

You should already have a board overlay file:

```
boards/<board>.overlay
```

How to do it...

1. **Define an alias in the overlay**
   Open boards/<board>.overlay and add:

```
/ {
    aliases {
        myadc = &adc1;
    };
};
```

   On an STM32 board, `myadc` points to `adc1`.

   On an NXP board, it would instead point to `lpadc0`:

```
/ {
    aliases {
        myadc = &lpadc0;
    };
};
```

2. **Use the alias in application code**
   In `src/main.c`

   Replace:

```
#define MY_ADC_NODE DT_NODELABEL(adc1)
```

   with:

```
#define MY_ADC_NODE DT_ALIAS(myadc)
```

3. **Rebuild and flash the application**
   The application output remains the same, but the code no longer depends on a board-specific node label.

How it works...

Aliases are defined inside the `special/aliases` node of the devicetree.

An alias does not create new hardware. Instead, it provides an application-facing name that references an existing node.

For example:

- One board may define its ADC as `adc1`
- Another board may define it as `lpadc0`

By using:

```
DT_ALIAS(myadc)
```

the application no longer depends on the original board-specific node label.

This improves portability because each board overlay can map the alias to the appropriate hardware node without changing the application source code.

There's more...

Devicetree also provides a special node `called/chosen`, which serves a similar purpose: it creates stable references for important system features.

However, `/chosen` is reserved for standard Zephyr subsystems and core functionality. For example:

```
/ {
    chosen {
        zephyr,console = &usart1;
    };
};
```

This redirects `printk()` output to a specific UART instance.

While aliases are meant for application-defined portability (such as `myadc`), `/chosen` is used for well-defined standard features such as console, flash, entropy, or SRAM.

See also

- Zephyr Project Documentation – Aliases and chosen nodes: `https://docs.zephyrproject.org/latest/build/dts/intro-syntax-structure.html#aliases-and-chosen-nodes`
- Zephyr Project Documentation – Devicetree chosen nodes: `https://docs.zephyrproject.org/latest/build/dts/api/api.html#devicetree-chosen-nodes`
- Zephyr Project Documentation – Devicetree API: `https://docs.zephyrproject.org/latest/build/dts/api/api.html`

Get this book's PDF copy, code bundle, and more

Scan the QR code (or go to `packtpub.com/unlock`). Search for this book by name, confirm the edition, and then follow the steps on the page.

UNLOCK NOW

Note: Have your invoice handy. Purchases made directly from the Packt website don't require an invoice.

5

Exploring Zephyr's Device Driver Model

Zephyr applications interact with hardware through standardized subsystem APIs rather than direct register access or vendor-specific hardware abstraction layers (HALs). This abstraction is built on top of Zephyr's device driver model and devicetree system, which together allow hardware to be described independently from application code while drivers are selected automatically during the build process.

In practice, directly accessing hardware registers or tightly coupling applications to specific peripherals quickly reduces portability and makes code difficult to maintain across boards and vendors. Without understanding how Zephyr represents devices, retrieves hardware information from devicetree, and dispatches operations to drivers, it becomes difficult to design reusable and scalable embedded applications.

In this chapter, you will explore Zephyr's device driver model and devicetree integration through a series of practical recipes. You will retrieve devices from Devicetree, control GPIOs, configure GPIO interrupts, and create a custom Devicetree binding for application-specific hardware descriptions.

Together, these recipes provide a practical approach to building portable Zephyr applications that interact with hardware through standardized and scalable abstractions.

In this chapter, we will cover the following recipes:

- Accessing hardware through the device model
- Configuring and using GPIOs
- Configuring and using GPIO interrupts
- Writing a custom Devicetree binding

Technical requirements

To follow the recipes in this chapter, you will need:

- A supported host machine
- Zephyr SDK or a compatible toolchain
- A supported development board (e.g., STM32, NXP) or a simulation environment
- Visual Studio Code with Workbench for Zephyr extension (optional but recommended)
- Preferably, the **OpenSmart Rich Shield**, which provides multiple peripherals used throughout the book, including a potentiometer
- Otherwise, a potentiometer connected to the Arduino A0 pin of the development board

The code files for this chapter are available in the downloadable code bundle under the Chapter 5 folder. You can download the code by following the instructions in the **Download the example code files** section in the *Preface*.

Accessing hardware through the device model

Applications should not access hardware registers or vendor-specific HALs directly. Doing so reduces portability and tightly couples the application to a specific microcontroller family.

Zephyr solves this problem through its device driver model. Applications interact with hardware using generic subsystem APIs, while the device model automatically dispatches operations to the correct hardware-specific driver.

In this recipe, you will retrieve an ADC device instance from Devicetree, perform an ADC read using the generic Zephyr ADC API, and observe how the driver model connects the generic API call to the underlying driver implementation.

Getting ready

Reuse the `my_adc` application created in the previous chapter.

Make sure (already done in chapter 4):

- Your overlay defines an alias named `myadc` that points to the ADC node
- The ADC node is enabled (`status = "okay"`)
- You already know which ADC channel you configured (for example, channel 3)

Remove everything from the main file.

How to do it...

1. **Enable the ADC driver**

   Add the following to `prj.conf`:

```
CONFIG_ADC=y
```

2. **Include required headers**

   Open `src/main.c` and include the ADC driver header alongside the devicetree headers from the previous chapter:

```
#include <zephyr/kernel.h>
#include <zephyr/drivers/adc.h>
```

3. **Retrieve the ADC node identifier and properties**

   For STM32 Nucleo H5 add:

```
#define MY_ADC_NODE DT_NODELABEL(adc1)
#define ADC_VREF_MV DT_PROP(MY_ADC_NODE, vref_mv)
#define MY_ADC_CH_NODE DT_CHILD(MY_ADC_NODE, channel_3)
```

   For FRDM MCXN236 add:

```
#define MY_ADC_NODE DT_NODELABEL(lpadc0)
#define ADC_VREF_MV DT_PROP(DT_CHILD(MY_ADC_NODE, channel_3),
zephyr_vref_mv)
#define MY_ADC_CH_NODE DT_CHILD(MY_ADC_NODE, channel_3)
```

4. **Retrieve the device and the channel configuration**

```
static const struct device *adc = DEVICE_DT_GET(MY_ADC_NODE);
static const struct adc_channel_cfg ch_cfg =
ADC_CHANNEL_CFG_DT(MY_ADC_CH_NODE);
```

   `DEVICE_DT_GET()` returns the struct device instance created by the driver model for the ADC. `ADC_CHANNEL_CFG_DT()` builds an `adc_channel_cfg` structure from the channel subnode's properties: gain, reference, acquisition time, channel ID, and driver-specific fields are filled in automatically.

5. **Write** `main()`

   Replace `main()` with:

```
int main(void)
{
    int err;
    uint32_t buf = 0;
    struct adc_sequence sequence = {
        .channels = BIT(ch_cfg.channel_id),
        .buffer=&buf,
        .buffer_size = sizeof(buf),
        .resolution = DT_PROP_OR(MY_ADC_CH_NODE, zephyr_resolution,
        12),
    };
    if (!device_is_ready(adc))
    {
        printk("ADC device not ready\n");
        return 0;
    }
    err = adc_channel_setup(adc,&ch_cfg);
    if (err < 0)
    {
        printk("Could not setup ADC channel (%d)\n", err);
        return 0;
    }
    while (1)
    {
        int32_t val_mv;
        buf = 0;
        err = adc_read(adc,&sequence);
        if (err < 0)
        {
            printk("Could not read ADC (%d)\n", err);
            k_sleep(K_MSEC(1000));
            continue;
        }
        if (ch_cfg.differential)
        {
            val_mv = (int32_t)((int16_t)buf);
        }
        else
```

```
        {
            val_mv = (int32_t)buf;
        }
        printk("ADC raw: %" PRId32, val_mv);
        err = adc_raw_to_millivolts(ADC_VREF_MV, ch_cfg.gain,
                                   sequence.resolution,&val_mv);
        if (err < 0)
        {
            printk(" (voltage conversion not available)\n");
        }
        else
        {
            printk(" Voltage: %" PRId32 " mV\n", val_mv);
        }
        k_sleep(K_MSEC(1000));
    }
    return 0;
}
```

6. **Rebuild and flash the application**

   You should see output similar to:

```
ADC raw: 842  Voltage: 370 mV
ADC raw: 1390  Voltage: 610 mV
ADC raw: 1327  Voltage: 583 mV
```

Turning a potentiometer connected to the ADC input changes the reported values.

How it works...

The generic ADC API is defined as:

```
int adc_read(const struct device *dev,
             const struct adc_sequence *sequence);
```

The key design point is that the API takes a `const struct device *`.

This pointer represents a device instance managed by Zephyr.

At runtime, struct device contains (among other fields) a pointer to a driver API table:

```
struct device {
    const char *name;
```

```
    const void *config;
    const void *api;
    void *data;
};
```

The api field is a void pointer because the generic device model must support many subsystems.

For ADC devices, `dev->api` points to a `struct adc_driver_api` instance.

Conceptually, the ADC subsystem uses a driver API (`adc_driver_api`) like:

```
typedef int (*adc_api_read)(const struct device *dev,
                const struct adc_sequence *sequence);

struct adc_driver_api {
    adc_api_read read;
};
```

So the generic `adc_read`(dev, seq) ends up calling the driver-specific read function through that function pointer:

```
((const struct adc_driver_api *)dev->api)->read(dev, seq)
```

On STM32, the ADC driver is implemented in:

```
drivers/adc/adc_stm32.c
```

That driver declares:

```
#define DT_DRV_COMPAT st_stm32_adc
```

This is the build-time link to devicetree nodes whose compatible is:

```
compatible = "st,stm32-adc";
```

All non-alphanumeric characters in the compatible string are converted to underscores when generating internal identifiers.

In the same driver file, the API table assigns the implementation of the read operation:

```
.read = adc_stm32_read
```

Swap the board for an NXP target and the same application code follows a different api pointer into `adc_mcux_lpadc_read()` instead. The application never names a vendor: the dispatch is wired up at link time from devicetree.

The full call chain is shown in the figure below:

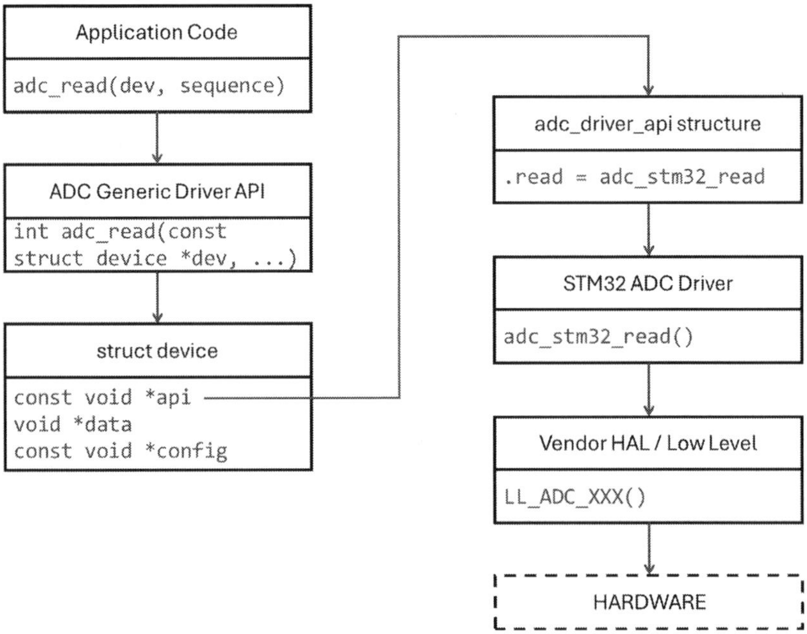

Figure 5.1 – Dispatch flow from generic ADC API to hardware-specific driver implementation

There's more...

In addition to the api pointer, struct device also contains two other important fields:

```
const void *config;
void *data;
```

These are internal driver structures.

- `config` typically holds constant configuration data derived from Devicetree (base address, IRQ number, clock settings, etc.).
- `data` holds the runtime state of the driver (buffers, flags, hardware status, internal context).

As an application developer, you normally do not access these fields directly. They are part of the driver's internal implementation.

However, they can be useful during debugging. When inspecting a struct device instance in a debugger, examining config and data can help you understand the internal state of the driver and verify that the hardware was initialized correctly.

In normal application code, you should always use the subsystem API (such as adc_read()), not the internal driver structures.

See also

- Zephyr Project Documentation – Device Driver Model: `https://docs.zephyrproject.org/latest/kernel/drivers/index.html`
- Zephyr Project Documentation – ADC API: `https://docs.zephyrproject.org/latest/hardware/peripherals/adc.html`
- Golioth Blog – Understanding Zephyr drivers and devices: `https://blog.golioth.io/how-to-write-a-zephyr-device-driver-with-a-custom-api/`

Configuring and using GPIOs

GPIOs are one of the most common peripherals used in embedded systems. Applications frequently need to control LEDs, read buttons, or interact with external hardware in a portable way without relying on vendor-specific APIs or hardcoded pin definitions.

Zephyr solves this problem through its GPIO subsystem and Devicetree integration. Hardware information such as the GPIO controller, pin number, and configuration flags can be retrieved directly from the Devicetree.

In this recipe, you will blink an LED using Zephyr's GPIO subsystem with the recommended `gpio_dt_spec` helper structure.

Getting ready

Create a new application named `mygpios` based on the `Hello World` template.

Use this minimal folder structure:

```
mygpios/
├── CMakeLists.txt
├── prj.conf
└── src/
    └── main.c
```

Enable GPIO support in `prj.conf`:

```
CONFIG_GPIO=y
```

Make sure your board provides a `led0` alias. Most boards do. You can check in the generated `zephyr.dts` file.

Search for an aliases section similar to:

```
aliases {
    led0 = ...;
};
```

If your board does not provide led0, you can connect an external LED to a GPIO pin and define it in an overlay.

For example, connect an LED and a resistor to a GPIO pin, then create:

```
boards/<your_board>.overlay
```

Add:

```
/ {
    leds {
        compatible = "gpio-leds";

        myled0: my_led {
            /* Set the actual used gpio group and pin */
            gpios = <&gpioa 5 GPIO_ACTIVE_HIGH>;
            label = "My LED";
        };
    };

    aliases {
        led0 = &myled0;
    };
};
```

Rebuild after adding the overlay.

How to do it...

1. **Retrieve the node identifier**

   Retrieve the LED node using the standard alias

   In `src/main.c`, add the required headers:

```
#include <zephyr/kernel.h>
#include <zephyr/drivers/gpio.h>
```

   Define the node identifier:

```
#define LED0_NODE DT_ALIAS(led0)
```

2. **Blinky using** `gpio_dt_spec`

   Use the `LED0_NODE` definition to create the `gpio_dt_spec` instance, add:

```
const struct gpio_dt_spec led =
    GPIO_DT_SPEC_GET(LED0_NODE, gpios);
```

3. **Configure and blink using** `*_dt` **helpers**

   Replace `main()` with:

```
int main(void)
{
    int ret;

    if (!device_is_ready(led.port)) {
        printk("GPIO device not ready: %s\n", led.port->name);
        return 0;
    }

    ret = gpio_pin_configure_dt(&led, GPIO_OUTPUT_INACTIVE);
    if (ret != 0) {
        printk("gpio_pin_configure_dt() failed: %d\n", ret);
        return 0;
    }

    while (1) {
        gpio_pin_toggle_dt(&led);
        k_sleep(K_MSEC(500));
```

```
        }

        return 0;
    }
```

Build and flash again. The behavior is identical, but the code is simpler.

How it works...

The led0 alias points to an LED node defined by the board (or by your overlay). That LED node contains a gpios property that describes the LED connection using three pieces of information:

- The GPIO controller (a devicetree node that becomes a struct device)
- The pin number
- The GPIO flags (active level and other attributes)

The `GPIO_DT_SPEC_GET()` reads all three from the devicetree and collects them into a single structure:

```
struct gpio_dt_spec {
    const struct device *port;
    gpio_pin_t pin;
    gpio_dt_flags_t dt_flags;
};
```

Grouping the controller, pin, and flags in one object matches the way GPIOs are described in devicetree, and lets the `*_dt` helper functions (`gpio_pin_configure_dt()`, `gpio_pin_toggle_dt()`, and so on) accept a single argument instead of three separate ones. The result is less boilerplate in application code and fewer opportunities to pass mismatched values.

For each `*_dt` function in the GPIO API, there is also a non-`*_dt` equivalent that takes the controller, pin, and flags as separate arguments. For example, `gpio_pin_configure_dt(&led, GPIO_OUTPUT_INACTIVE)` corresponds to `gpio_pin_configure(led.port, led.pin, GPIO_OUTPUT_INACTIVE)`. The non-`*_dt` versions still work, but they require you to unpack the spec yourself and re-apply the flags every time, which is exactly the boilerplate `gpio_dt_spec` is designed to eliminate. Prefer the `*_dt` helpers whenever you have a spec.

There's more...

`gpio_dt_spec` is one example of Zephyr's Devicetree "spec" helpers. Similar helpers exist for other subsystems such as ADC and I2C.

When a subsystem provides a *_dt_spec type, it is usually best practice to use it. It reduces boilerplate, keeps the code consistent, and improves portability across boards.

See also

- Zephyr Project Documentation – GPIO API: `https://docs.zephyrproject.org/latest/hardware/peripherals/gpio.html`
- Zephyr Project Documentation – GPIO Devicetree helpers: `https://docs.zephyrproject.org/latest/build/dts/api/api.html`
- Golioth Blog – Zephyr for Hardware Engineers: GPIO: `https://blog.golioth.io/zephyr-for-hardware-engineers-gpio/`

Configuring and using GPIO interrupts

Polling a GPIO pin in a loop is inefficient and wastes CPU time. A better approach is to react to hardware events using interrupts.

In this recipe, you will configure a GPIO pin as a button input and use interrupts to toggle an LED when the button is pressed.

Zephyr already provides a standard Devicetree binding for buttons called gpio-keys. It is a generic Devicetree representation for one or more push buttons connected to GPIO pins. Boards that support buttons typically expose them through gpio-keys and provide a convenient alias such as `sw0`.

You will add a button to the same project used in the previous recipe. When the button is pressed, your application will toggle the LED defined by led0.

Getting ready

Reuse the same `mygpios` project from the previous recipe.

Most boards already define a user button using the sw0 alias. You can check in the generated `zephyr.dts`.

Search for:

```
aliases {
    sw0 = ...;
};
```

If your board does not provide `sw0`, you can connect an external push button to a GPIO pin and define it in an overlay.

For example, connect:

- One side of the button to a GPIO pin
- The other side to GND

Then, in `boards/<your_board>.overlay`, add a gpio-keys node and an alias:

```
/ {
    keys {
        compatible = "gpio-keys";

        sw0: button_0 {
            gpios = <&gpioa 0 GPIO_ACTIVE_LOW>;
            label = "User Button";
        };
    };

    aliases {
        sw0 = &sw0;
    };
};
```

Rebuild after adding the overlay.

How to do it...

1. **Retrieve the LED and button using aliases**

   In `src/main.c`, add the required headers:

```
#include <zephyr/kernel.h>
#include <zephyr/drivers/gpio.h>
```

   Define the node identifiers:

```
#define LED0_NODE DT_ALIAS(led0)
#define SW0_NODE  DT_ALIAS(sw0)
```

   Retrieve both as `gpio_dt_spec`:

```
const struct gpio_dt_spec led = GPIO_DT_SPEC_GET(LED0_NODE, gpios);
const struct gpio_dt_spec button = GPIO_DT_SPEC_GET(SW0_NODE, gpios);
```

2. **Define a callback**

   Add a `gpio_callback` object and a state variable:

```
struct gpio_callback button_cb;
bool led_on;
```

   Add the callback function:

```
void button_pressed(const struct device *dev,
                    struct gpio_callback *cb,
                    uint32_t pins)
{
    ARG_UNUSED(dev);
    ARG_UNUSED(cb);
    ARG_UNUSED(pins);

    led_on = !led_on;
    gpio_pin_set_dt(&led, (int)led_on);
}
```

3. **Configure the { XE "GPIOs:using" }GPIOs and enable { XE "GPIOs:configuring" } the interrupt**

   Replace `main()` with:

```
int main(void)
{
    int ret;

    if (!device_is_ready(led.port) || !device_is_ready(button.port)) {
        printk("GPIO device not ready\n");
        return 0;
    }

    ret = gpio_pin_configure_dt(&led, GPIO_OUTPUT_INACTIVE);
    if (ret != 0) {
        printk("Failed to configure LED: %d\n", ret);
        return 0;
    }

    ret = gpio_pin_configure_dt(&button, GPIO_INPUT);
```

```
    if (ret != 0) {
        printk("Failed to configure button: %d\n", ret);
        return 0;
    }

    ret = gpio_pin_interrupt_configure_dt(&button,
GPIO_INT_EDGE_TO_ACTIVE);
    if (ret != 0) {
        printk("Failed to configure interrupt: %d\n", ret);
        return 0;
    }

    gpio_init_callback(&button_cb, button_pressed, BIT(button.pin));
    gpio_add_callback(button.port, &button_cb);

    printk("Press the button to toggle the LED\n");

    while (1) {
        k_sleep(K_FOREVER);
    }

    return 0;
}
```

Build and flash. Each button press should toggle the LED.

How it works...

The button is described in Devicetree using the gpio-keys binding. This binding describes the electrical connection of the button using a `gpios` property, including the controller, pin number, and flags (such as pull-up and active-low).

When you call `gpio_pin_interrupt_configure_dt()`, Zephyr configures the GPIO controller hardware to generate an interrupt when the pin transitions to the active state.

When the button is pressed, the interrupt fires and the GPIO driver calls the registered callback function. The callback runs in interrupt context, so it should be short and non-blocking. In this recipe, the callback simply toggles a boolean state and sets the LED output accordingly.

There's more...

Buttons can also be handled in polling mode using the gpio-keys binding. This is useful when you want a simple design and you do not need interrupt-based behavior.

To enable polling and software debouncing, update the button node in your overlay and add these properties:

```
/ {
    my_buttons: buttons {
        compatible = "gpio-keys";
        debounce-interval-ms = <30>;
        polling-mode;

        sw0: button_0 {
            gpios = <&gpioa 0 GPIO_ACTIVE_LOW>;
            label = "User Button";
        };
    };
};
```

In polling mode, you typically read the button using `gpio_pin_get_dt(&button)` inside a loop and detect a press by comparing the current state with the previous one.

See also

- Zephyr Project Documentation – GPIO API: `https://docs.zephyrproject.org/latest/hardware/peripherals/gpio.html`
- Zephyr Project Documentation – gpio-keys binding: `https://docs.zephyrproject.org/latest/build/dts/api/bindings/input/gpio-keys.html`

Writing a custom Devicetree binding

When working with custom hardware, you may need to describe new devices that are not covered by existing Devicetree bindings. Without a binding, Zephyr cannot validate or interpret the node correctly.

In this recipe, you will create a custom Devicetree binding, define its properties, and use it from an application.

You will define one required GPIO property for an LED and one optional delay property. Then, you will write a small application that blinks the LED using the delay from the Devicetree.

Getting ready

Create a new application from the `Hello World` template and name it `mytotoled`:

```
mytotoled/
├── CMakeLists.txt
├── prj.conf
├── dts/
│   └── bindings/
│       └── toto-led.yaml
├── boards/
│   └── <your_board>.overlay
└── src/
    └── main.c
```

Enable GPIO in `prj.conf`:

```
CONFIG_GPIO=y
```

This recipe assumes you have one LED available:

- Either an onboard LED
- Or an external LED connected to a GPIO pin

How to do it...

1. **Create the custom Devicetree binding**

   Create `dts/bindings/toto-led.yaml` with the following content:

```
compatible: "toto-led"

description: Example custom node describing a single LED GPIO and a blink
delay.

properties:
  toto-gpios:
    type: phandle-array
    required: true
    description: My LED GPIO.

  my-delay-ms:
    type: int
```

```
    required: false
    default: 500
    description: Blink delay in milliseconds.
```

2. **Add a Devicetree node using the binding**
   Create `boards/<your_board>.overlay` and add a node that uses the binding. The `toto-gpios` property must point to an actual LED pin on your board.

   You can either reuse an onboard LED or connect an external one:

   On STM32 Nucleo H5, the onboard green LED is on `gpiob 0`:

   ```
   toto-gpios = <&gpiob 0 GPIO_ACTIVE_HIGH>;
   ```

   On FRDM-MCXN236, the onboard red LED is on `gpio4` 18:

   ```
   toto-gpios = <&gpio4 18 GPIO_ACTIVE_HIGH>;
   ```

   Otherwise, connect an external LED to any free GPIO pin and set `toto-gpios` to match that pin.

   ```
   / {
       toto_led0: toto_led0 {
           compatible = "toto-led";
           toto-gpios = <&gpiob 0 GPIO_ACTIVE_HIGH>;
           my-delay-ms = <1000>; /* Optional, equals 500 by default */
       };

       aliases {
           mytotoled = &toto_led0;
       };
   };
   ```

   Rebuild after adding the overlay.

3. **Use the node from application code**
   In `src/main.c`, add the following code:

   ```
   #include <zephyr/kernel.h>
   #include <zephyr/drivers/gpio.h>
   ```

```
#define TOTO_NODE DT_ALIAS(mytotoled)

const struct gpio_dt_spec led = GPIO_DT_SPEC_GET(TOTO_NODE, toto_gpios);

#define BLINK_DELAY_MS DT_PROP(TOTO_NODE, my_delay_ms)

int main(void)
{
    int ret;

    if (!device_is_ready(led.port)) {
        printk("LED GPIO device not ready\n");
        return 0;
    }

    ret = gpio_pin_configure_dt(&led, GPIO_OUTPUT_INACTIVE);
    if (ret != 0) {
        printk("Failed to configure LED: %d\n", ret);
        return 0;
    }

    printk("Blinking with delay %d ms\n", BLINK_DELAY_MS);

    while (1) {
        gpio_pin_toggle_dt(&led);
        k_sleep(K_MSEC(BLINK_DELAY_MS));
    }

    return 0;
}
```

Build and flash. The LED should blink using the configured delay.

How it works...

The YAML binding (`toto-led.yaml`) defines the expected structure of any node that uses it:

```
compatible = "toto-led";
```

It declares that:

- `toto-gpios` is required and must be a GPIO phandle-array, which means you must specify the GPIO controller, the pin number, and the pin configuration (for example `<&gpioa 5 GPIO_ACTIVE_HIGH>`)
- `my-delay-ms` is optional and defaults to 500

During the build, Zephyr merges devicetree sources and overlays, then validates nodes against their bindings. If you do not define `toto-gpios`, the build fails with a devicetree/binding error because the property is marked as required.

In the application, `GPIO_DT_SPEC_GET()` reads `toto-gpios` and creates a `gpio_dt_spec` structure (controller device, pin, and flags). `DT_PROP()` reads `my-delay-ms` as a compile-time constant. If `my-delay-ms` is not provided in the overlay, the default value from the binding (500) is used.

There's more...

Compatible strings typically follow the format: "company,feature"

Zephyr checks the company part against the vendor-prefixes list. If the prefix is not listed, you usually get a warning during the devicetree processing stage.

If your device is not manufactured by a specific hardware vendor, do not invent a vendor prefix. Vendor prefixes are not mandatory parts of compatible properties, and compatibles should not include them unless they refer to an actual vendor.

See also

- Zephyr Project Documentation – Devicetree bindings: `https://docs.zephyrproject.org/latest/build/dts/bindings.html`
- Zephyr Project Documentation – Custom bindings: `https://docs.zephyrproject.org/latest/build/dts/howtos.html`
- Interrupt Blog – Practical Zephyr Devicetree: `https://interrupt.memfault.com/blog/practical_zephyr_dt`

Get this book's PDF copy, code bundle, and more

Scan the QR code (or go to `packtpub.com/unlock`). Search for this book by name, confirm the edition, and then follow the steps on the page.

UNLOCK NOW

Note: Have your invoice handy. Purchases made directly from the Packt website don't require an invoice.

6

System Initialization and Multithreading

Modern embedded applications rarely follow a sequential execution path. Instead, they must handle multiple tasks concurrently, such as reading sensors, processing data, responding to events, and communicating with peripherals, often within strict timing constraints. Managing these activities in a single loop quickly becomes difficult to scale and maintain.

Without a clear understanding of concurrency, it becomes difficult to structure applications correctly, manage interactions between different execution paths, or ensure predictable system behavior. As systems grow in complexity, poorly managed concurrency can lead to unresponsive applications, hard-to-debug issues, and inefficient use of system resources.

In this chapter, you will explore how Zephyr executes application code concurrently and how threads interact with each other during runtime. Understanding how threads are created, scheduled, and coordinated is essential when building responsive and maintainable Zephyr applications. The recipes in this chapter introduce the core mechanisms used to structure concurrent applications in Zephyr, starting with how code executes during system startup, followed by creating and managing threads, delaying and blocking execution, exchanging data between threads, and synchronizing their behavior. You will also explore how workqueues allow tasks to be deferred and processed asynchronously without creating additional threads.

Together, these recipes provide a structured approach to designing and managing multithreaded Zephyr applications in real-world scenarios.

In this chapter, we will cover the following recipes:

- Executing code automatically at startup
- Creating threads in Zephyr
- Delaying and blocking threads in Zephyr
- Exchanging data between threads in Zephyr
- Synchronizing threads in Zephyr
- Deferring work using Zephyr workqueues

Technical requirements

To follow the recipes in this chapter, you will need:

- A supported host machine
- Zephyr SDK or a compatible toolchain
- A supported development board (e.g., STM32, NXP) or a simulation environment
- Visual Studio Code with Workbench for Zephyr extension (optional but recommended)

The code files for this chapter are available in the downloadable code bundle under the Chapter 6 folder. You can download the code by following the instructions in the **Download the example code files** section in the *Preface*.

Executing code automatically at startup

When a Zephyr system starts, many important operations happen before `main()` runs. Drivers, kernel services, and application components may all need initialization at specific stages of the boot process. Without understanding this sequence, it can be difficult to place startup code correctly.

In this recipe, you will use `SYS_INIT()` callbacks to observe Zephyr's initialization levels, verify execution order at runtime, and understand how code can run automatically before `main()`.

Getting ready

You need:

- A working Zephyr installation (west workspace set up)
- We will use the Zephyr Hello World sample as the starting point
- Serial console output working

How to do it...

1. **Create the application structure**

   Open the Create New Application panel in Workbench for Zephyr.

   - Select your target board
   - Select Sample project: `Hello World`
   - In Project Name, enter: `init_toto`
   - Finish the wizard

2. **Register initialization callbacks**

   Replace the contents of `src/main.c` with the following:

```
#include <zephyr/kernel.h>

int post_kernel_cb(void) {
    printk("[POST_KERNEL:10] Called automatically by the kernel\n");
    return 0;
}

int post_kernel_cb2(void) {
    printk("[POST_KERNEL:20] Called after priority 10\n");
    return 0;
}

int application_cb(void) {
    printk("[APPLICATION] Called automatically before main()\n");
    return 0;
}

SYS_INIT(post_kernel_cb,  POST_KERNEL,  10);
SYS_INIT(post_kernel_cb2, POST_KERNEL,  20);
SYS_INIT(application_cb,  APPLICATION,  10);

int main(void)
{
    printk("[main] running\n");
    return 0;
}
```

None of these functions are called explicitly. Registering them with `SYS_INIT()` is sufficient, the kernel automatically invokes them during startup.

3. **Build and run the application**

   Build and flash the application. Open your serial terminal. You should see:

```
[POST_KERNEL:10] Called automatically by the kernel
[POST_KERNEL:20] Called after priority 10
*** Booting Zephyr OS build v4.4.1 ***
[APPLICATION] Called automatically before main()
[main] running
```

   The output confirms the execution order: `POST_KERNEL` callbacks run first in priority order, followed by the boot banner, then the `APPLICATION` callback, and finally `main()`.

How it works...

Zephyr divides system startup into ordered initialization levels such as EARLY, PRE_KERNEL_1, PRE_KERNEL_2, POST_KERNEL, and APPLICATION. Each level represents a stage where specific components are initialized.

The `SYS_INIT`(function, level, priority) macro registers a callback that runs automatically during one of these stages. Within the same level, lower priority numbers execute first.

In this recipe, two callbacks are registered at the `POST_KERNEL` level with different priorities. Because `post_kernel_cb` uses priority 10 and `post_kernel_cb2` uses priority 20, the first function executes before the second. The `application_cb` function is registered at the APPLICATION level, so it runs later in the boot sequence, just before `main()` is executed.

The internal implementation of this mechanism can be found in the Zephyr source tree in `kernel/init.c`. The kernel entry point begins in `z_cstart()`, which orchestrates the initialization sequence. The function repeatedly calls `z_sys_init_run_level()` to execute all registered `SYS_INIT` callbacks for each initialization level.

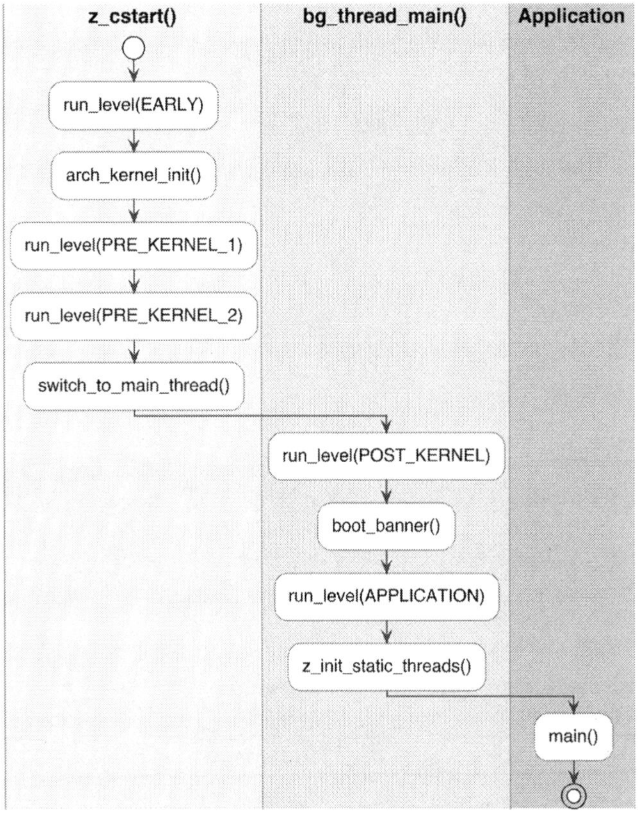

Figure 6.1 – Zephyr startup flow from z_cstart() to main()

Figure 6.1 illustrates this sequence. After the CPU starts executing `z_cstart()`, Zephyr runs the EARLY and PRE_KERNEL_1 and 2 initialization stages. These stages prepare low-level architecture components and kernel structures before the scheduler is active. Once these steps are complete, execution switches to the main thread.

From there, the POST_KERNEL initialization level runs, followed by the boot banner, and finally the APPLICATION initialization level. After all application-level initialization callbacks have executed, Zephyr initializes statically defined threads and then calls `main()`, where the application logic begins.

Drivers register their initialization functions in either the PRE_KERNEL_X or POST_KERNEL stages depending on whether they require kernel services. Low-level hardware drivers that must run before the kernel is fully operational often use PRE_KERNEL levels, while drivers that rely on kernel features should initialize in POST_KERNEL.

For application development, initialization code is most typically registered at the APPLICATION level. This allows application components to run setup code automatically before `main()` begins executing.

Many Zephyr subsystems also rely on this mechanism. For example, when the Bluetooth subsystem is enabled, several initialization functions are automatically registered and executed during startup. Because these functions run in the context of the main thread, they can affect the required stack size of the main thread, which sometimes must be increased using `CONFIG_MAIN_STACK_SIZE` depending on the enabled subsystems.

There's more...

To inspect the final call order produced by the linker, Zephyr provides the `initlevels` build target:

```
west build -t initlevels
```

This command prints all initialization entries in the exact order they will run at boot, including both `SYS_INIT` callbacks and device initialization functions.

This is useful when debugging initialization dependencies or when trying to understand why a driver or subsystem starts before or after another component.

It is also worth noting that the `main()` callback is optional in Zephyr. The function is defined as a weak symbol in the kernel. If the application does not provide its own implementation, the default version simply returns. In practice, an application can perform all initialization work using `SYS_INIT()` callbacks registered at the APPLICATION level, without defining a `main()` function at all.

See also

- Zephyr Project Documentation, `SYS_INIT()`: `https://docs.zephyrproject.org/latest/doxygen/html/group__sys__init.html`
- Zephyr Project: RTOS Start-up and Initialization Flow: `https://osselc21.sched.com/event/lAUk/virtual-zephyr-project-rtos-start-up-and-initialization-flow-david-leach-nxp`
- Startup function `z_cstart()` and main thread `bg_thread_main()`: `https://github.com/zephyrproject-rtos/zephyr/blob/main/kernel/init.c`

Creating threads in Zephyr

Many embedded applications need different tasks to run independently, such as reading sensors, handling communication, or updating outputs. Trying to execute everything in a single loop quickly becomes difficult to maintain and can reduce responsiveness. Threads solve this by allowing multiple execution paths to run under the scheduler.

In this recipe, you will create threads dynamically with `k_thread_create()`, define threads statically with `K_THREAD_DEFINE()`, and compare how priorities affect execution order.

Getting ready

Start from the standard Hello World template and create a new application named `multithreaded_toto`.

How to do it...

1. **Define stacks and priorities**

   Open `src/main.c` and add the includes and stack definitions:

```
#include<zephyr/kernel.h>

#define THREAD_A_STACK_SIZE 512
#define THREAD_A_PRIORITY   2

#define THREAD_B_STACK_SIZE 512
#define THREAD_B_PRIORITY   10

K_THREAD_STACK_DEFINE(thread_a_stack, THREAD_A_STACK_SIZE);

struct k_thread thread_a_data;
k_tid_t thread_a_id;
```

2. **Create the entry functions**

   Add the two thread entry functions. Each thread prints its priority value:

```
void thread_a_entry(void *p1, void *p2, void *p3) {
        printk("thread_A: I am at high priority, value: %d \n",
                k_thread_priority_get(k_current_get()));


}
```

```
void thread_b_entry(void *p1, void *p2, void *p3) {
        printk("thread_B: I am at low priority, value: %d \n",
                k_thread_priority_get(k_current_get()));
}
```

3. **Create thread A inside main**

   Create thread A at runtime in the `main()` callback:

```
int main(void)
{
    thread_a_id = k_thread_create(&thread_a_data,
                thread_a_stack, K_THREAD_STACK_SIZEOF(thread_a_stack),
                thread_a_entry, NULL, NULL, NULL,
                THREAD_A_PRIORITY, 0, K_NO_WAIT);

    return 0;
}
```

4. **Statically define and initialize thread B**

   Use `K_THREAD_DEFINE()` macro to allocate and create thread B automatically at startup, outside any function:

```
K_THREAD_DEFINE(thread_b_id, THREAD_B_STACK_SIZE,
                thread_b_entry, NULL, NULL, NULL,
                THREAD_B_PRIORITY, 0, 0);
```

5. **Build and run the application**

   Build and flash the application. Open a serial terminal. Both threads run once and exit. Since `thread_A` has a higher priority (lower numeric value), it runs first. After it finishes execution, `thread_B` runs and prints its message:

```
thread_A: I am at high priority, value: 2
thread_B: I am at low priority, value: 10
```

How it works...

A thread created with `k_thread_create()` is instantiated explicitly at runtime. The application provides the thread control structure, stack memory, entry function, and priority. Once created, the thread becomes runnable immediately because `K_NO_WAIT` was used.

A thread declared with `K_THREAD_DEFINE()` is allocated statically at build time and started automatically during system initialization. This approach is useful for threads that should always exist when the application starts and avoids manually creating and exposing the stack memory buffer.

Zephyr schedules runnable threads according to priority. Lower numeric values represent higher priority. In this example, `thread_A` uses priority 2 and `thread_B` uses priority 10, so `thread_A` runs first.

Figure 6.2 shows the Zephyr priority model, where lower numeric values represent higher scheduling priority.

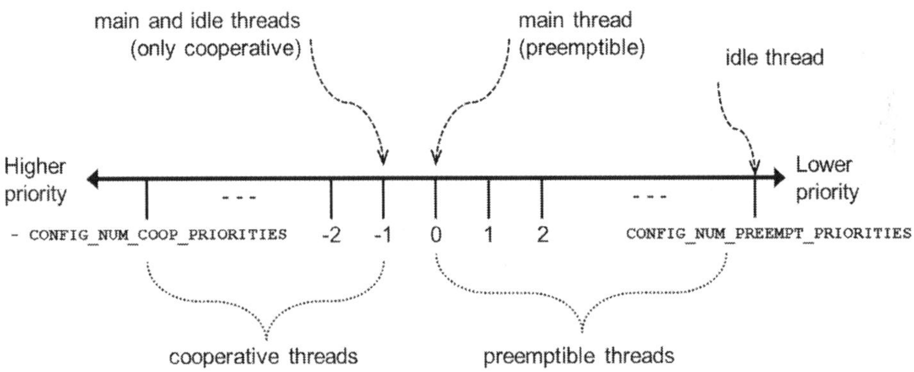

Figure 6.2 – Zephyr thread priority ordering

There's more...

Zephyr supports three scheduling policies, and you choose the right one based on what your thread needs to do:

- **Preemptive scheduling:** Use for most application threads. The scheduler can interrupt a lower-priority preemptive thread the moment a higher-priority one becomes ready. Priority values 0 and above opt into this policy.
- **Cooperative scheduling:** Use for threads that must never be interrupted mid-sequence, such as a critical state machine or a protocol handler. A cooperative thread holds the CPU until it explicitly yields or blocks. Assign a negative priority value to opt into this policy.
- **Timeslicing:** Use when multiple threads share the same priority and you want fair round-robin execution between them. Enable it with `CONFIG_TIMESLICING=y` and configure the slice duration with `CONFIG_TIMESLICE_SIZE`. Without timeslicing, two equal-priority preemptive threads only switch when one blocks.

See also

- Zephyr Project Documentation, Threads: https://docs.zephyrproject.org/latest/kernel/services/threads/index.html
- Zephyr Project Documentation, Scheduling: https://docs.zephyrproject.org/latest/kernel/services/scheduling/index.html
- Digikey, Multithreading: https://www.digikey.nl/en/maker/tutorials/2025/introduction-to-zephyr-part-8-multithreading

Delaying and blocking threads in Zephyr

Threads rarely run continuously without interruption. In real applications, a thread often needs to pause for a specific time, wait for another task to complete, or remain blocked until an event occurs. Efficient waiting is essential to avoid wasting CPU time.

In this recipe, you will delay threads with `k_sleep()`, suspend and resume threads explicitly, and understand how Zephyr represents timeout values with `k_timeout_t`.

Getting ready

Reuse the same `multithreaded_toto` project created in the previous recipe.

How to do it...

1. **Suspend the high-priority thread**
   The high-priority thread prints a message, then suspends itself. `k_thread_suspend()` takes a thread ID as its argument. The ID of the currently executing thread can be obtained using `k_current_get()`:

```
void thread_a_entry(void *p1, void *p2, void *p3) {
    printk("thread_A: running, about to suspend\n");
    k_thread_suspend(k_current_get());
    printk("thread_A: resumed, exiting\n");
}
```

2. **Delay the low-priority thread**

The low-priority thread prints a message, sleeps for 500 ms using `k_sleep()` with a `K_MSEC()` timeout. After waking, it prints another message and resumes the suspended high-priority thread:

```
void thread_b_entry(void *p1, void *p2, void *p3) {
    printk("thread_B: running, sleeping for 500ms\n");
    k_sleep(K_MSEC(500));
    printk("thread_B: woke up after 500ms, resuming thread_A\n");
    k_thread_resume(thread_a_id);
}
```

3. **Build and run the application**

   Build and flash the application. You should see output similar to:

```
thread_A: running, about to suspend
thread_B: running, sleeping for 500ms
thread_B: woke up after 500ms, resuming thread_A
thread_A: resumed, exiting
```

How it works...

`k_timeout_t` is an opaque struct type. You should construct it using one of Zephyr's timeout macros. The call `k_sleep(K_MSEC(500))` passes a `k_timeout_t` value to the kernel that encodes 500 milliseconds. Internally, the kernel converts that value to system ticks at the point it enters the timeout queue. The precision of the actual delay is bounded by the tick rate of the system clock.

`K_MSEC(500)` produces a `k_timeout_t` representing a relative delay of 500 milliseconds from the moment the kernel receives it. Other constructors follow the same pattern and accept the same type wherever a timeout is expected in the API.

`k_thread_suspend()` is different from sleeping. A sleeping thread will wake up automatically once the timeout expires. A suspended thread does not have an associated timeout. It is removed from the scheduler's ready queue and remains blocked indefinitely until another thread explicitly calls `k_thread_resume()`.

There's more...

Zephyr uses the `k_timeout_t` type across many kernel APIs, not only with `k_sleep()`. Instead of writing raw tick values directly, applications typically construct timeout values using helper macros. This keeps the code readable and portable across systems with different tick configurations.

Common timeout constructors include the following:

Macro	Unit
`K_NSEC(n)`, `K_USEC(n)`, `K_MSEC(n)`, `K_SECONDS(n)`, `K_MINUTES(n)`, `K_HOURS(n)`	Nanoseconds, Microseconds, Milliseconds, Seconds, Minutes, Hours
`K_TICKS(n)`	System ticks
`K_NO_WAIT`	No wait (immediate)
`K_FOREVER`	Wait indefinitely

Table 6.1 – Common Zephyr timeout constructor macros

These timeout values are widely used throughout the Zephyr kernel. For example, the same `k_timeout_t` type is accepted by APIs for sleeping, waiting on semaphores, receiving from queues, and many other blocking operations.

Explicit suspension using `k_thread_suspend()` and `k_thread_resume()` is rarely used directly in typical application code. In most designs, threads block while waiting for events through kernel synchronization primitives such as semaphores, FIFOs, or message queues. When these APIs are called with a timeout like `K_FOREVER`, the kernel internally performs the equivalent of an indefinite suspension: the thread is removed from the scheduler's ready queue and remains blocked until the awaited event occurs. In practice, this means that applications usually rely on blocking APIs with `K_FOREVER` rather than manually suspending and resuming threads.

See also

- Zephyr Project Documentation, Kernel Timing: `https://docs.zephyrproject.org/latest/kernel/services/timing/clocks.html`

Exchanging data between threads in Zephyr

Threads often need to pass information to each other. A sensor thread may produce measurements, a communication thread may receive packets, or a worker thread may generate results that another task must process. Sharing data directly can quickly become unsafe or difficult to coordinate.

In this recipe, you will use a Zephyr message queue to transfer structured data safely between a producer thread and a consumer thread, while allowing both threads to run independently.

Getting ready

You can start from the `multithreaded_toto` application created in the previous recipe or create a copy of it to keep this example separate.

How to do it...

1. **Define the message type**

   Add the following structure at the top of `src/main.c` to describe the data exchanged between threads:

```
struct toto_msg {
    int tata;
    int titi;
};
```

2. **Declare the message queue**

   Declare the message queue at global scope using `K_MSGQ_DEFINE()`. This macro defines both the queue and its internal ring buffer. The message queue can hold up to 3 messages, each the size of struct `toto_msg`, aligned on a 4-byte boundary:

```
K_MSGQ_DEFINE(toto_msgq, sizeof(struct toto_msg), 3, 4);
```

3. **Create the producer entry function**

   The high-priority thread creates a message and sends it to the message queue. `k_msgq_put()` copies the message into its internal ring buffer. `K_NO_WAIT` means the call returns immediately if the message queue is full rather than blocking:

```
void thread_a_entry(void *p1, void *p2, void *p3)
{
   struct toto_msg msg = {
       .tata = 42,
       .titi = 100,
   };

   k_msgq_put(&toto_msgq,&msg,K_NO_WAIT);
   printk("thread_A: sent tata=%d titi=%d\n", msg.tata, msg.titi);
}
```

4. **Create the consumer entry function**

   The low-priority thread blocks on `k_msgq_get()` using `K_FOREVER`, meaning it waits indefinitely until a message becomes available. When a message arrives, it will be copied into the local `msg` variable:

```
void thread_b_entry(void *p1, void *p2, void *p3){
    struct toto_msg msg;

    k_msgq_get(&toto_msgq,&msg,K_FOREVER);
    printk("thread_B: received tata=%d titi=%d\n", msg.tata, msg.titi);
}
```

5. **Build and run the application**

   Build and flash the application. You should see output similar to:

```
thread_A: sent tata=42 titi=100
thread_B: received tata=42 titi=100
```

How it works...

A message queue acts as a fixed-size ring buffer shared between threads. When the producer calls `k_msgq_put()`, the kernel copies, byte per byte, the entire struct `toto_msg` into the next available slot in the queue.

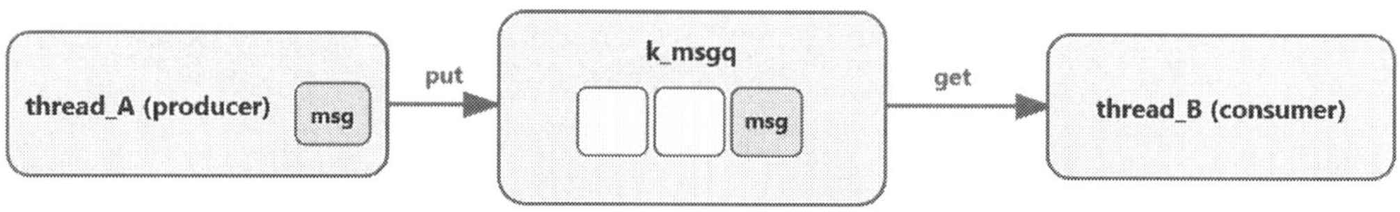

Figure 6.3 – Producer-consumer communication using a Zephyr message queue

If the queue is full, the behavior depends on the timeout argument. With `K_NO_WAIT` the call returns immediately. With a blocking timeout, the thread waits until space becomes available.

When the consumer calls `k_msgq_get()` with `K_FOREVER`, it blocks if the queue is empty. As soon as the producer inserts a message, the kernel wakes the consumer, copies the message from the ring buffer into the local variable, and frees the queue slot.

There's more...

A FIFO is another option when threads need to exchange data items. Unlike a message queue, which copies data into an internal fixed-size buffer, a FIFO passes pointers to dynamically allocated items. This allows variable-sized messages and avoids copying.

Because the kernel uses the first word of each item to maintain its internal linked list, your structure must reserve that space explicitly:

```
struct toto_msg {
    void *reserved;
    int   tata;
    int   titi;
};
```

You declare the FIFO with `K_FIFO_DEFINE()` and must allocate each message yourself before putting it in:

```
K_FIFO_DEFINE(toto_fifo);

void thread_a_entry(void *p1, void *p2, void *p3) {
    struct toto_msg *msg = k_malloc(sizeof(struct toto_msg));

    msg->tata = 42;
    msg->titi = 100;
    k_fifo_put(&toto_fifo,msg);
}

void thread_b_entry(void *p1, void *p2, void *p3) {
    struct toto_msg *msg = k_fifo_get(&toto_fifo,K_FOREVER);

    printk("received tata=%d titi=%d\n", msg->tata, msg->titi);
    k_free(msg);
}
```

Because this example uses `k_malloc()`, the system heap must be enabled. For example, add the following to `prj.conf`:

```
CONFIG_HEAP_MEM_POOL_SIZE=512
```

Unlike the message queue, the FIFO holds no fixed upper limit on the number of items. The trade-off is that you are responsible for allocating and freeing each message, and the reserved first word is mandatory, omitting it causes undefined behavior as the kernel will overwrite the first field of your struct.

See also

- Zephyr Project Documentation, Message Queues: `https://docs.zephyrproject.org/latest/kernel/services/data_passing/message_queues.html`
- Zephyr Project Documentation, FIFOs: `https://docs.zephyrproject.org/latest/kernel/services/data_passing/fifos.html`

Synchronizing threads in Zephyr

Threads often depend on each other. One task may need to wait until another finishes some work, receives data, or signals that an event has occurred. Without synchronization, threads can run in the wrong order or waste CPU time checking conditions repeatedly.

In this recipe, you will coordinate two threads using a Zephyr semaphore and learn when the polling API `k_poll()` becomes useful for waiting on multiple events.

Getting ready

You can start from the `multithreaded_toto` application created in the previous recipe, or create a copy of it to keep this example separate.

How to do it...

1. **Define the semaphore**
   Add the following semaphore definition at global scope in `src/main.c`:

   ```
   K_SEM_DEFINE(sync_sem, 0, 1);
   ```

   The initial count is set to 0, meaning that any thread attempting to take the semaphore will block until another thread releases it.

2. **Create the waiting thread**
   The high-priority thread waits for the semaphore. Because the semaphore count is initially zero, the thread blocks until another thread signals it.

   ```
   void thread_a_entry(void *p1, void *p2, void *p3)
   {
       printk("thread_A: waiting for semaphore\n");

       k_sem_take(&sync_sem,K_FOREVER);
   ```

```
    printk("thread_A: semaphore received, continuing\n");
}
```

3. **Create the signaling thread**
   The second thread performs some work and then releases the semaphore to wake the waiting thread.

```
void thread_b_entry(void *p1, void *p2, void *p3)
{
    printk("thread_B: doing work\n");

    k_sleep(K_SECONDS(1));

    printk("thread_B: signaling semaphore\n");

    k_sem_give(&sync_sem);
}
```

4. **Build and run the application**
   Build and flash the application. You should see output similar to:

```
thread_A: waiting for semaphore
thread_B: doing work
thread_B: signaling semaphore
thread_A: semaphore received, continuing
```

How it works...

A semaphore stores an internal counter managed by the kernel. When a thread calls `k_sem_take()`, the kernel checks the current value of that counter.

If the counter is greater than zero, the kernel decrements it and the thread continues immediately. If the counter is zero, the thread cannot proceed and is removed from the scheduler ready queue until the semaphore becomes available.

When another thread calls `k_sem_give()`, the counter is incremented. If one or more threads are waiting, the kernel wakes one waiting thread and allows it to continue.

In this example, `thread_A` blocks because the semaphore starts at zero. After `thread_B` finishes its simulated work, it calls `k_sem_give()`, which wakes `thread_A`.

Semaphores are commonly used to signal events between threads, or between an interrupt handler and a worker thread.

There's more...

In more complex systems, a thread may need to wait for multiple types of events simultaneously. Zephyr provides the polling API (k_poll()) to support this pattern.

Using k_poll(), a thread can wait on several kernel objects at the same time, such as: semaphores, message queues, and FIFOs...

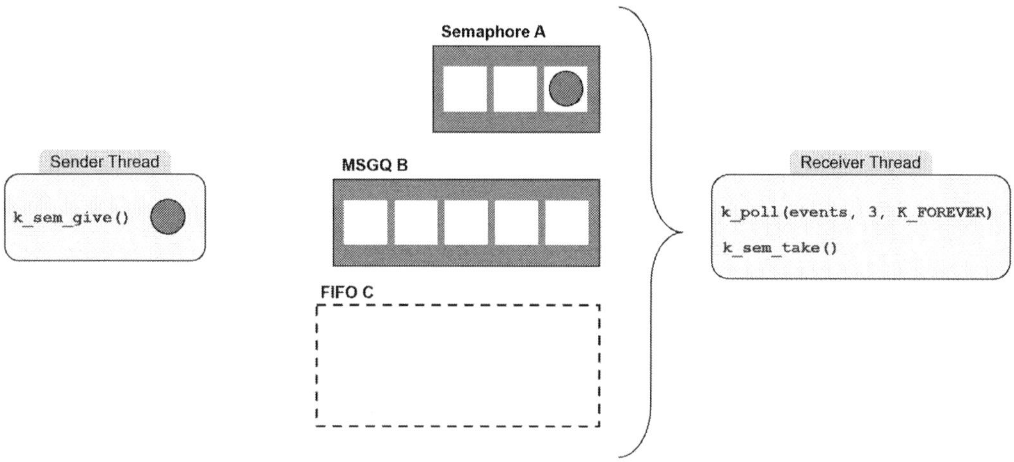

Figure 6.4 – Polling multiple synchronization objects

For example, a worker thread may wait for a semaphore that signals a hardware event, a message queue that delivers processed data, or a FIFO that carries dynamically allocated messages. The thread blocks until any of these objects become ready.

This allows a single thread to react to multiple event sources without continuously polling them in a loop, making the system more efficient and easier to design.

See also

- Zephyr Project Documentation, Semaphores: `https://docs.zephyrproject.org/latest/kernel/services/synchronization/semaphores.html`
- Zephyr Project Documentation, Polling API: `https://docs.zephyrproject.org/latest/kernel/services/polling.html`
- Common multithreading problems and their fixes: `https://www.zephyrproject.org/common-multithreading-problems-and-their-fixes/`

Deferring work using Zephyr workqueues

Some tasks should not be processed immediately in the context where they are detected. An interrupt handler, callback, or time-sensitive thread may need to react quickly and postpone heavier processing until later. Creating a dedicated thread for every small task is often unnecessary.

In this recipe, you will use Zephyr workqueues to defer processing, submit custom work items asynchronously, and recover application data inside the work handler with `CONTAINER_OF()`.

Getting ready

Start from the standard `Hello World` template and create a new application named `workqueue_toto`.

Build once to confirm the `workqueue_toto` application works correctly.

How to do it...

1. **Define a custom structure containing a work item**
   Open `src/main.c` and replace its contents with the following code:

```
#include <zephyr/kernel.h>

struct toto_operation {
    int event_id;
    int value;
    struct k_work work;
};
```

   The struct `k_work` field is embedded inside the application-defined structure. This allows each work item to carry its own application data.

2. **Create the work handler**
   Add a work handler function that receives a pointer to struct `k_work`. Use `CONTAINER_OF()` to recover the enclosing `toto_operation` structure:

```
void toto_work_handler(struct k_work *work)
{
    struct toto_operation *op =
        CONTAINER_OF(work, struct toto_operation, work);
```

```
    printk("workqueue: processing event_id=%d value=%d\n",
           op->event_id, op->value);
}
```

3. **Declare work items**

   Create three statically allocated operations at global scope:

```
struct toto_operation op1 = { .event_id = 1, .value = 10};

struct toto_operation op2 = { .event_id = 2, .value = 20};

struct toto_operation op3 = { .event_id = 3, .value = 30};
```

4. **Initialize and submit the work items**

   In `main()`, initialize the embedded work items and submit them to the system workqueue. Submit two operations back-to-back, then delay briefly before submitting the third one:

```
int main(void)
{
    printk("main: initializing work items\n");

    k_work_init(&op1.work,toto_work_handler);
    k_work_init(&op2.work,toto_work_handler);
    k_work_init(&op3.work,toto_work_handler);

    printk("main: submitting op1\n");
    k_work_submit(&op1.work);

    printk("main: submitting op2\n");
    k_work_submit(&op2.work);

    k_sleep(K_MSEC(500));

    printk("main: submitting op3\n");
    k_work_submit(&op3.work);
```

```
    return 0;
}
```

5. **Build and run the application**
   Build and flash the application. Open your serial terminal. You should see output similar to:

```
main: initializing work items
main: submitting op1
main: submitting op2
workqueue: processing event_id=1 value=10
workqueue: processing event_id=2 value=20
main: submitting op3
workqueue: processing event_id=3 value=30
```

How it works...

A work item is represented by struct `k_work`. In this recipe, struct `k_work` is embedded inside a larger application-defined structure named `toto_operation`. This pattern is common in Zephyr because it allows the work handler to access additional application data associated with the work item.

When `main()` calls `k_work_submit()`, the work item is placed in the system workqueue. The system workqueue is created by default in Zephyr. Its thread executes the submitted work items one by one.

The work handler receives only a pointer to struct `k_work`. Since the work item is embedded inside struct `toto_operation`, the handler uses `CONTAINER_OF()` to recover the pointer to the enclosing structure. This gives access to `event_id` and value without needing global variables or separate lookup logic.

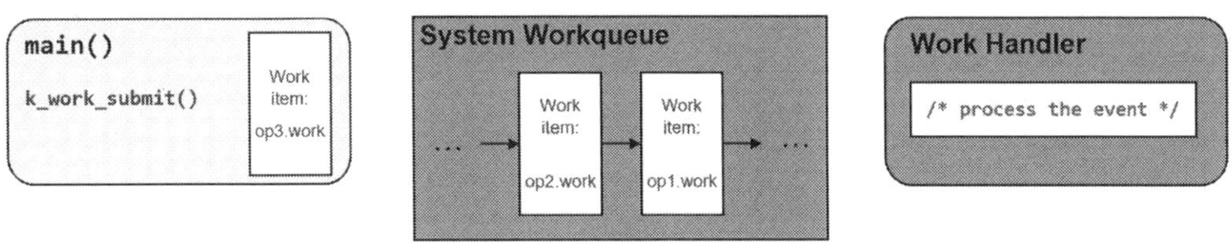

Figure 6.5 – Work items processed by the system workqueue

Submitting two work items consecutively shows that multiple operations can be queued for later execution.

There's more...

The work items in this recipe are processed by the system workqueue, which is sufficient for many applications. Zephyr also allows applications to create custom workqueues when more control is needed.

A custom workqueue can be useful when deferred work should run at a different priority, be isolated from the system workqueue, or handle heavier processing separately. In that case, the application defines its own struct `k_work_q`, stack, and worker thread, then starts the queue with `k_work_queue_start()`.

Workqueues are widely used in Zephyr drivers and subsystems because they provide a simple way to defer processing out of time-sensitive contexts without creating additional dedicated threads for every task.

See also

- Zephyr Project Documentation, Workqueues: `https://docs.zephyrproject.org/latest/kernel/services/threads/workqueue.html`
- Zephyr's Work Queues: Explained in Simple Words: `https://www.linkedin.com/pulse/zephyrs-work-queues-explained-simple-words-roy-jamil-eapye`
- Using Custom Work Queues for Sensor Readings in Zephyr: `https://blog.golioth.io/using-custom-work-queues-for-sensor-readings-in-zephyr/`

Get this book's PDF copy, code bundle, and more

Scan the QR code (or go to `packtpub.com/unlock`). Search for this book by name, confirm the edition, and then follow the steps on the page.

UNLOCK NOW

Note: Have your invoice handy. Purchases made directly from the Packt website don't require an invoice.

7

Working with Zephyr Subsystems

Modern Zephyr applications rely on reusable subsystems that provide standardized APIs for common functionality such as randomness, logging, command interfaces, messaging, and runtime tracing. These subsystems allow applications to remain portable and modular while the underlying implementation can change depending on the hardware or configuration.

In practice, directly interacting with hardware-specific drivers or building custom solutions for common tasks quickly leads to code that is difficult to maintain, reuse, and scale across different boards. Without a clear understanding of how subsystems abstract these details, it can be challenging to select the right interfaces, configure them correctly, and integrate them into larger applications.

In this chapter, you will explore several commonly used Zephyr subsystems and learn how they simplify application development. Through a series of practical recipes, you will generate random values using the **Random** subsystem, configure and manage structured logging, interact with your application through the **Shell** subsystem, broadcast messages using the **zbus** subsystem, and analyze system behavior using tracing tools.

Together, these recipes provide a practical approach to using Zephyr subsystems to build modular, portable, and observable embedded applications.

In this chapter, we will cover the following recipes:

- Generate random numbers using the Random subsystem
- Configuring and managing logging in Zephyr
- Interacting with your application using shell
- Broadcasting messages using Zephyr zbus
- Tracing and visualizing Zephyr runtime behavior

Technical requirements

To follow the recipes in this chapter, you will need:

- A supported host machine
- Zephyr SDK or a compatible toolchain
- A supported development board (e.g., STM32, NXP) or a simulation environment
- Visual Studio Code with Workbench for Zephyr extension (optional but recommended)

The code files for this chapter are available in the downloadable code bundle under the Chapter 7 folder. You can download the code by following the instructions in the **Download the example code files** section in the *Preface*.

Generate random numbers using the Random subsystem

When developing portable embedded applications, directly calling hardware-specific drivers quickly makes the code harder to reuse across different boards. A better approach is to rely on Zephyr subsystems, which provide stable high-level APIs while selecting the correct backend through configuration.

In this recipe, you will use the Random subsystem as a practical example to understand how Zephyr subsystems connect application code, Kconfig options, Devicetree information, and backend drivers.

Getting ready

Start from the standard `Hello World` template and create a new application named `myrandom`.

Use the following minimal structure:

```
myrandom/
├── CMakeLists.txt
├── prj.conf
└── src/
    └── main.c
```

Build once to confirm your `Hello World` application works.

You will now extend this project to use the Random subsystem.

How to do it...

1. **Check whether the board provides an entropy device**
   Open the generated devicetree file `zephyr.dts`.

   Search for a chosen node similar to:

```
/ {
    chosen {
        zephyr,entropy = &rng;
    };
};
```

   The `zephyr,entropy` property selects the entropy device used by the Random subsystem.

   If your board defines this property, the Random subsystem can use the hardware entropy source automatically.

   If your board does not define an entropy device by default, that is acceptable for experimentation. In that case, you can enable the test random generator. However, the test generator produces predictable values and must not be used in production or release builds.

2. **Select a random backend**
   Open `prj.conf` and enable one of the following options:

   To use a hardware-backed generator:

```
CONFIG_ENTROPY_DEVICE_RANDOM_GENERATOR=y
```

   This option works only when the selected board provides a supported entropy device through Devicetree and the corresponding driver is enabled.

   For testing only:

```
CONFIG_TEST_RANDOM_GENERATOR=y
```

3. **Replace the application code**
   Now open `src/main.c` and replace its contents with:

```
#include <zephyr/kernel.h>
#include <zephyr/random/random.h>
```

```
int main(void)
{
    while (1) {
        uint32_t value = sys_rand32_get();
        printk("Random value: %u\n", value);
        k_sleep(K_MSEC(1000));
    }

    return 0;
}
```

4. **Build and flash the application**

   Build the firmware and flash it to the board similar to what we have done in previous recipes.

   If a hardware entropy source is available, you will see non-deterministic values printed every second.

   If the test generator is enabled, the values will follow a predictable pattern.

How it works...

The function `sys_rand32_get()` is part of Zephyr's Random subsystem. Your application calls a stable API without knowing how the random value is actually produced.

At build time, Zephyr selects the backend implementation using Kconfig options and Devicetree information. If `CONFIG_ENTROPY_DEVICE_RANDOM_GENERATOR` is enabled and the board defines a supported `zephyr,entropy` device, the subsystem uses that hardware source. Other configurations may select software or test backends.

This is the key idea behind subsystems in Zephyr: application code depends on a common API, while the build system connects that API to the appropriate implementation for the selected target.

There's more...

The Random subsystem also provides a cryptographically secure API:

```
int sys_csrand_get(void *dst, size_t len);
```

This function generates cryptographically secure random data suitable for encryption, authentication, and security tokens.

Secure random generation may rely on hardware entropy devices or a **CSPRNG** implementation. Whenever randomness affects security, `sys_csrand_get()` should be used instead of `sys_rand32_get()`.

The same subsystem design principles apply: your application calls a stable API, and Zephyr selects the appropriate secure backend based on the configuration.

See also

- Zephyr Project Documentation, Random APIs: `https://docs.zephyrproject.org/latest/doxygen/html/group__random__api.html`
- Zephyr Project Documentation, Entropy Drivers: `https://docs.zephyrproject.org/latest/hardware/peripherals/entropy.html`
- Digikey, Kconfig Tutorial: `https://www.digikey.com/en/maker/tutorials/2025/introduction-to-zephyr-part-3-kconfig-tutorial`

Configuring and managing logging in Zephyr

As applications grow, simple `printk()` statements quickly become difficult to manage. Messages have no severity levels, filtering is limited, and uncontrolled console output can increase code size and reduce performance. Zephyr solves this with a structured logging subsystem.

In this recipe, you will enable logging, replace `printk()` with logging macros, configure log levels, and observe how filtering changes the runtime output.

Getting ready

In this recipe, you will extend the existing `myrandom` application and replace `printk()` with structured logging. You will configure log levels, observe filtering behavior, and manage log output using Kconfig options.

How to do it...

1. **Enable the logging subsystem**
   Open `prj.conf` and enable the logging subsystem:

   ```
   CONFIG_LOG=y
   CONFIG_LOG_DEFAULT_LEVEL=3
   ```

   The default level 3 corresponds to informational messages.

2. **Replace `printk()` with logging macros**

   Now open `src/main.c` and replace `printk()` usage with logging macros:

```
#include <zephyr/kernel.h>
#include <zephyr/logging/log.h>
#include <zephyr/random/random.h>

LOG_MODULE_REGISTER(myrandom_app);

int main(void)
{
    while (1) {
        uint32_t value = sys_rand32_get();

        LOG_DBG("Generated value: %u", value);
        LOG_INF("Random value: %u", value);
        LOG_WRN("Example warning message");
        LOG_ERR("Example error message");

        k_sleep(K_MSEC(1000));
    }

    return 0;
}
```

3. **Build and flash the application**

   Build the firmware and flash it to the board.

   With `CONFIG_LOG_DEFAULT_LEVEL=3`, you should see:

```
[00:00:00.003,000] <inf> myrandom_app: Random value: 2758
[00:00:00.003,000] <wrn> myrandom_app: Example warning message
[00:00:00.003,000] <err> myrandom_app: Example error message
```

   You should not see `LOG_DBG()` messages because the debug level is lower than the configured threshold.

4. **Enable debug messages**

   Change `prj.conf` to:

```
CONFIG_LOG_DEFAULT_LEVEL=4
```

Rebuild and flash again. You should now also see:

```
[00:00:00.005,000] <dbg> myrandom_app: main: Generated value: 2758
```

Other default debug messaging might also appear, for example:

```
[00:00:00.000,000] <dbg> os: setup_thread_stack: stack 0x20001548 for
thread 0x20000628: obj_size=1024 buf_start=0x20001548  buf_size 1024
stack_ptr=0x20001948
```

How it works...

Each logging macro sends a message with an associated severity level:

- LOG_DBG – Debug
- LOG_INF – Informational
- LOG_WRN – Warning
- LOG_ERR – Error

The line:

```
LOG_MODULE_REGISTER(myrandom_app);
```

registers a logging module for this source file. The module name is myrandom_app, and it allows the logging subsystem to identify the source of log messages and apply module-specific filtering rules. This becomes especially useful in larger applications, where logs from different source files or components need to be distinguished.

The logging subsystem uses Kconfig to determine which severity levels are compiled into the application. Messages below the configured level may be completely removed during compilation.

This means that disabling debug logging does not only hide output, it can reduce code size and runtime overhead.

The logging system processes messages and forwards them to a backend. By default, logs are typically sent to the console backend using UART.

Because logging is implemented as a subsystem, your application code remains independent of how logs are transported or displayed.

There's more...

The logging subsystem can be configured in several ways depending on how logs should be processed and where they should be sent.

By default, Zephyr uses deferred logging (`CONFIG_LOG_MODE_DEFERRED`). In this mode, log messages are placed in a buffer and processed later by a dedicated logger thread (`CONFIG_LOG_PROCESS_THREAD`). This minimizes the impact on the application because formatting and output are performed outside the main execution path.

Logs are typically sent to the UART console backend (`CONFIG_LOG_BACKEND_UART`), which prints messages on the serial terminal. For better readability, you can enable options such as `CONFIG_LOG_BACKEND_SHOW_COLOR` to highlight warnings and errors, and `CONFIG_LOG_BACKEND_FORMAT_TIMESTAMP` to print human-readable timestamps. If the logging buffer becomes full, enabling `CONFIG_LOG_MODE_OVERFLOW` allows the logger to discard the oldest messages to make space for new ones.

Besides UART, Zephyr supports other backends. One useful example is the Bluetooth logging backend, which allows logs to be transmitted over Bluetooth Low Energy. This can be useful when no physical debug interface is available. The Bluetooth backend uses the Nordic UART Service (NUS) so it can be viewed using tools such as the nRF Toolbox mobile application.

These options allow the logging system to be adapted to different environments, whether logs are collected through a serial console, a debug probe, or a wireless connection.

See also

- Zephyr Project Documentation, Logging subsystem: `https://docs.zephyrproject.org/latest/services/logging/index.html`
- Debugging Zephyr for Beginners: `printk()` and the Logging Subsystem: `https://blog.golioth.io/debugging-zephyr-for-beginners-printk-and-the-logging-subsystem/`
- Using the logger module: `https://academy.nordicsemi.com/courses/nrf-connect-sdk-fundamentals/lessons/lesson-3-printing-messages-to-console-and-logging/topic/exercise-2-4/`

Interacting with your application using shell

Testing embedded features through rebuild-and-flash cycles can be slow and inefficient. Developers often need a faster way to inspect system behavior, trigger functions, and adjust runtime settings while the device is running. Zephyr solves this with the Shell subsystem, which provides an interactive command-line interface.

In this recipe, you will enable the shell subsystem, access it through the serial terminal, and extend it with a custom command that generates random values.

Getting ready

In this recipe, you will extend the existing `myrandom` application created in the previous recipes.

The board must provide a serial console so the shell can be accessed through a terminal.

The shell communicates through a UART device defined in Devicetree. If the board contains multiple UART peripherals, you may need to specify which one should host the shell interface using the chosen node.

How to do it...

1. **Enable the shell subsystem**

   Open `prj.conf` and enable the shell and its serial backend:

```
CONFIG_SHELL=y
CONFIG_SHELL_BACKEND_SERIAL=y
```

   The shell will use the UART backend.

2. **Select the UART used by the shell**

   You probably do not need to do this step as most existing boards already have a chosen serial peripheral selected by default.

   If the board has multiple UART peripherals, update the Devicetree chosen node:

```
/ {
    chosen {
        zephyr,shell-uart = &usart3; /* chose your specific uart if needed
*/
    };
};
```

   This configuration tells Zephyr to attach the shell interface to the USART3 peripheral.

3. **Include the required headers**

   Open `src/main.c` and include the required headers:

```
#include <zephyr/kernel.h>
#include <zephyr/shell/shell.h>
#include <zephyr/random/random.h>
```

4. **Create the command handler**

   Define a function that generates a random value when the command is executed:

```
static int cmd_random_get(const struct shell *shell, size_t argc, char
**argv)
{
    uint32_t value = sys_rand32_get();

    shell_print(shell, "Random value: %u", value);

    return 0;
}
```

5. **Create a help subcommand**

   Add a second command that prints usage information:

```
static int cmd_random_help(const struct shell *shell, size_t argc, char
**argv)
{
    shell_print(shell, "myrandom get   : generate a random value");
    shell_print(shell, "myrandom help  : show this message");

    return 0;
}
```

6. **Define and register command**

   Create a subcommand set containing both commands:

```
SHELL_STATIC_SUBCMD_SET_CREATE(sub_myrandom,
    SHELL_CMD(get, NULL, "Generate a random value", cmd_random_get),
    SHELL_CMD(help, NULL, "Show command help", cmd_random_help),
    SHELL_SUBCMD_SET_END
);
```

> **Note**
>
> Defining and registering commands should be done in the `main.c` at the global level, outside any function like `main()`.

Register the root command with the shell subsystem:

```
SHELL_CMD_REGISTER(myrandom, &sub_myrandom, "Random number commands",
NULL);
```

7. **Build and run the application**

   Build and flash the firmware, then open a serial terminal connected to the selected UART.

   Use the help command to list the available commands:

```
uart:~$ help
```

   You should see output similar to:

```
Available commands:
  myrandom   : Random number commands
  help       : Prints the help message
  history    : Command history
```

   You can now interact with the application:

```
uart:~$ myrandom get
Random value: 14523423
uart:~$ myrandom help
myrandom get    : generate a random value
myrandom help   : show this message
```

How it works...

When the shell subsystem is enabled, Zephyr initializes an interactive command-line interface during system startup. The shell listens for user input from the configured backend, which in this recipe is the UART serial interface.

Commands are registered using macros such as `SHELL_CMD_REGISTER()`. Each command is associated with a handler function that is executed when the command is invoked.

Subcommands are grouped using `SHELL_STATIC_SUBCMD_SET_CREATE()`, this structure allows commands to be organized in a hierarchy, similar to commands in a traditional CLI. In this example, the root command `myrandom` contains two subcommands: get and help.

When a command is entered in the terminal, the shell parses the input and calls the associated handler function. The handler receives the shell instance and command arguments, and can send output back to the user using functions such as `shell_print()`.

This mechanism allows applications to expose runtime functionality through a simple and extensible command interface.

There's more...

The Zephyr shell subsystem supports several communication backends. While UART is the most commonly used interface during development, other backends are available, including USB, Telnet, Bluetooth, **SEGGER RTT**, and **MQTT**.

The shell is particularly useful during development and prototyping because it allows developers to inspect system behavior, trigger operations, and test features interactively. However, it is not common to keep the shell enabled in production firmware. Removing it from release builds reduces memory usage and avoids exposing an interactive control interface on deployed devices.

See also

- Zephyr Project Documentation, Shell subsystem: `https://docs.zephyrproject.org/latest/services/shell/index.html`
- Zephyr: How to use Shells for Prototyping & Debugging Notecard Applications: `https://dev.blues.io/blog/zephyr-use-shells-for-prototyping/`
- How to add custom shell commands in Zephyr: `https://blog.golioth.io/how-to-add-custom-shell-commands-in-zephyr/`

Broadcasting messages using Zephyr zbus

As applications grow, directly coupling threads and modules through function calls can make the system harder to maintain and extend. Components often need a cleaner way to exchange data and broadcast events without necessarily knowing who produces the information or which parts of the system will consume it. Zephyr solves this with the zbus subsystem, which provides channel-based publish/subscribe communication.

In this recipe, you will create a zbus channel, broadcast values to multiple observers, validate messages, and observe how listeners and subscriber threads react to the same publication.

Getting ready

You can start from the shell-based `myrandom` application created in the previous recipe, or create a copy of it to keep this example separate.

You need:

- A working Zephyr application structure
- Logging enabled
- Shell already enabled if you want to keep the `myrandom` command
- Serial console output working

If you continue from the previous shell recipe, keep the shell configuration and the existing `myrandom` command.

How to do it...

1. **Enable Zbus**
   Zbus kconfig option should be enabled in `prj.conf`:

   ```
   CONFIG_ZBUS=y
   ```

2. **Include the required headers**
   Open `src/main.c` and include the required headers:

   ```
   #include <zephyr/kernel.h>
   #include <zephyr/logging/log.h>
   #include <zephyr/random/random.h>
   #include <zephyr/shell/shell.h>
   #include <zephyr/zbus/zbus.h>
   ```

3. **Create a validator**
   Add a validator. In this recipe, only values below 100 are accepted:

   ```
   static bool random_validator(const void *msg, size_t msg_size)
   {
       const int *value = msg;

       ARG_UNUSED(msg_size);

       return *value < 100;
   }
   ```

4. **Define the channel and the observers**

   Add a listener callback. A listener runs immediately when the channel is published, which makes it useful for reporting activity:

```
static void random_listener_cb(const struct zbus_channel *chan)
{
    const int *value = zbus_chan_const_msg(chan);
    LOG_INF("listener noticed publication: %d", *value);
}

ZBUS_LISTENER_DEFINE(random_listener, random_listener_cb);
ZBUS_SUBSCRIBER_DEFINE(random_sub_a, 4);
ZBUS_SUBSCRIBER_DEFINE(random_sub_b, 4);
```

   Now define the channel. This recipe uses a simple integer message and attaches the validator and all observers when the channel is created:

```
ZBUS_CHAN_DEFINE(random_chan,
         int,
         random_validator,
         NULL,
         ZBUS_OBSERVERS(random_listener, random_sub_a, random_sub_b),
         ZBUS_MSG_INIT(0));
```

5. **Create the subscriber threads**

   Add two thread entry functions. Each subscriber waits for a publication, reads the current channel value, and reports it with LOG_INF():

```
static void subscriber_a_thread(void *p1, void *p2, void *p3)
{
    ARG_UNUSED(p1);
    ARG_UNUSED(p2);
    ARG_UNUSED(p3);

    const struct zbus_channel *chan;
    int value;

    while (1) {
        if (zbus_sub_wait(&random_sub_a,&chan, K_FOREVER) == 0) {
            zbus_chan_read(chan, &value, K_MSEC(200));
            LOG_INF("subscriber A received: %d", value);
```

```
        }
    }
}

static void subscriber_b_thread(void *p1, void *p2, void *p3)
{
    ARG_UNUSED(p1);
    ARG_UNUSED(p2);
    ARG_UNUSED(p3);

    const struct zbus_channel *chan;
    int value;

    while (1) {
        if (zbus_sub_wait(&random_sub_b,&chan, K_FOREVER) == 0) {
            zbus_chan_read(chan, &value, K_MSEC(200));
            LOG_INF("subscriber B received: %d", value);
        }
    }
}
```

Create the subscriber threads using `K_THREAD_DEFINE`:

```
K_THREAD_DEFINE(sub_a_tid, 1024, subscriber_a_thread,
        NULL, NULL, NULL, 5, 0, 0);

K_THREAD_DEFINE(sub_b_tid, 1024, subscriber_b_thread,
        NULL, NULL, NULL, 5, 0, 0);
```

6. **Publish values**

   Use the shell-based `myrandom` application style from the previous recipe, you should add a publish subcommand.

   To do that, add a shell handler such as:

```
static int cmd_random_publish(const struct shell *shell, size_t argc, char
**argv)
{
    ARG_UNUSED(argc); ARG_UNUSED(argv);

    int value = sys_rand32_get() % 120;
```

```
    if (zbus_chan_pub(&random_chan, &value, K_MSEC(100)) == 0) {
        shell_print(shell, "Published value: %d", value);
    } else {
        shell_print(shell, "Rejected value: %d", value);
    }

    return 0;
}
```

Then add the publish subcommand to the existing `myrandom` shell command set.

7. **Build and validate**

   Build and flash the application, then open the serial terminal.

   Now publish a value:

```
uart:~$ myrandom publish
Published value: 42
<inf> myrandom: listener noticed publication: 42
<inf> myrandom: Subscriber A received: 42
<inf> myrandom: Subscriber B received: 42
```

   If the published value is rejected by the validator, you should see:

```
<inf> myrandom: Rejected value: 108
```

How it works...

A zbus channel stores a single message instance and a list of observers attached to that channel. In this recipe, the message type is an integer and the observers are one listener and two subscribers.

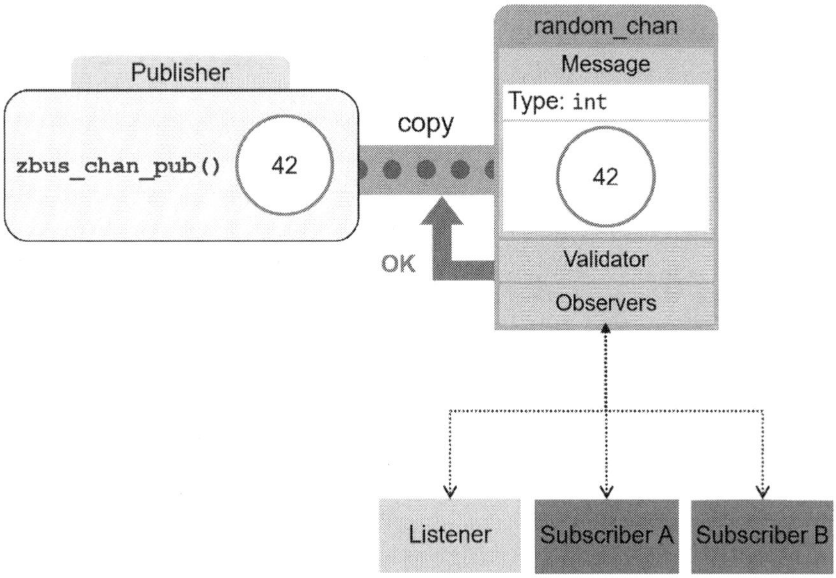

Figure 7.1 – zbus message publish flow

When the publisher calls `zbus_chan_pub()`, the first step performed by zbus is running the validator associated with the channel. If the validator accepts the value, the channel message is updated and all observers are notified. If the validator rejects the value, the publication fails, and the previous message stored in the channel remains unchanged.

The listener runs immediately in the context of the publisher during the publication process. Because it executes directly during the publish operation, it is useful for quickly reporting that a publication occurred.

Subscribers behave differently. Each subscriber runs in its own thread and waits for a notification using `zbus_sub_wait()`. When a channel is published, zbus notifies subscribers through an internal message queue associated with each subscriber. The message queue does not contain the actual message data. Instead, it contains the address of the channel that generated the event. This is an important detail because it is sometimes misunderstood: the subscriber does not receive the message itself through the message queue, it only receives a notification that the channel has new data.

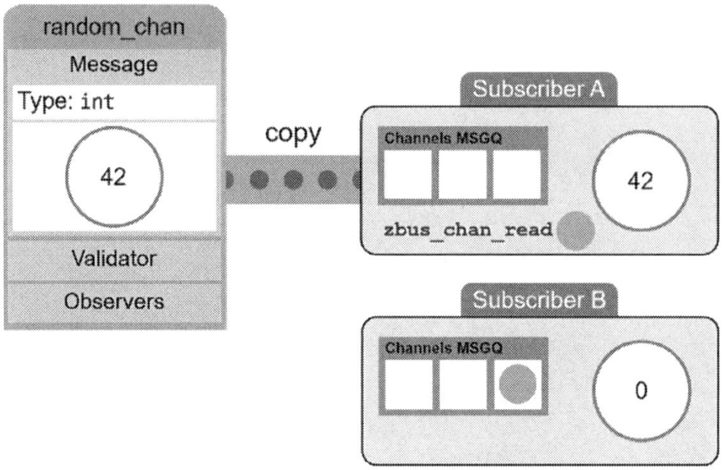

Figure 7.2 – zbus subscriber processing flow

After being notified, the subscriber thread reads the current message from the channel using `zbus_chan_read()` and processes it in its own execution context.

One limitation of regular subscribers is that they do not keep a private copy of every publication. They only receive a notification that the channel changed and then read the current value from the channel. If the same channel is published multiple times before a subscriber has time to process the earlier notification, the older value may be overwritten and effectively lost.

There's more…

In this recipe, the channel message is a simple integer, but zbus channels can also carry structures. In real applications, a structure is often more useful because it can group several fields in one message, such as a value, a timestamp, and a status.

If losing earlier publications is not acceptable, you can use message subscribers instead of regular subscribers. A message subscriber keeps a copy of each published message for each subscriber, instead of only notifying that the shared channel value changed. This avoids replacing earlier messages before a subscriber reads them, but it requires more memory because each subscriber needs storage for its own copied messages.

See also

- Zephyr Project Documentation, zbus: `https://docs.zephyrproject.org/latest/services/zbus/index.html`
- How to use Zephyr zbus to Communicate Between Threads: `https://blog.golioth.io/how-to-use-zephyr-zbus-to-communicate-between-threads/`
- Zephyr's ZBus Feature with Rodrigo Peixoto: `https://www.zephyrproject.org/zephyrs-zbus-feature/`

Tracing and visualizing Zephyr runtime behavior

Logs and breakpoints are useful, but they often do not show how the system behaves over time. Tracing tools record runtime events on a timeline and help identify issues such as thread scheduling patterns, CPU load, blocking or waiting threads, interrupt activity, context-switch frequency, and timing interactions between tasks.

This type of visibility is especially useful when debugging responsiveness problems, performance bottlenecks, or unexpected system behavior.

In this recipe, you will enable tracing in a Zephyr application, capture runtime activity into a RAM buffer, extract the trace data, and analyze thread execution using Percepio View for Zephyr.

Getting ready

- In this recipe, you will use the Zephyr philosophers sample application, make sure to use the sample under samples/philosophers. This sample implements the classic dining philosophers problem, where several threads represent philosophers competing for shared resources (forks). It is a good example for tracing because it involves multiple threads that frequently change state.
- You will also need to install Percepio View for Zephyr. This is a free and lightweight trace viewer compatible with **Tracealyzer** recordings. Download it from: `https://traceviewer.io/`
- Percepio View uses the same trace format as Tracealyzer. If you want the full Tracealyzer tool, you can purchase or request a trial license from: `https://percepio.com/tracealyzer/`
- In this recipe, the development environment used is Workbench for Zephyr running on Visual Studio Code.
- Make sure you can build and debug Zephyr applications successfully before continuing.

How to do it...

1. **Create the sample application**

   Create a new application named `my_philosophers` based on the Zephyr philosophers sample located in `samples/philosophers`, you need to select the "philosophers" sample instead of the `hello_world` project.

2. **Configure Percepio tracing**

   Open `prj.conf` and enable tracing with the Percepio TraceRecorder library by adding the following configuration lines:

   ```
   CONFIG_TRACING=y
   CONFIG_PERCEPIO_TRACERECORDER=y
   ```

   Enable snapshot mode using a ring buffer stored in RAM:

   ```
   CONFIG_PERCEPIO_TRC_CFG_STREAM_PORT_RINGBUFFER=y
   CONFIG_PERCEPIO_TRC_CFG_STREAM_PORT_RINGBUFFER_SIZE=8192
   ```

   Snapshot mode stores the trace data in a memory buffer instead of streaming it through a debug interface, the size of the buffer can be increased if needed, this will allow you to store trace data for a longer period.

   The tracing configuration can also be customized using the configuration GUI.

   Open the configuration interface and navigate to:

   ```
   (Top) -> Modules -> percepio -> TraceRecorder
   ```

   You should see configuration options similar to the following:

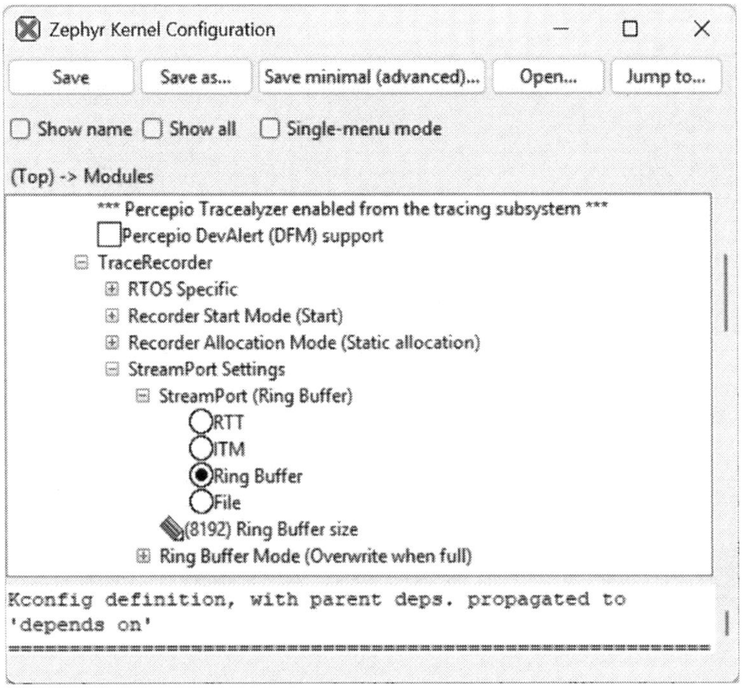

Figure 7.3 – TraceRecorder configuration in guiconfig

These options allow you to configure the stream port, buffer size, and recorder behavior.

3. **Build and run the application**

   Build the project and start a debugging session using Workbench for Zephyr.

   Let the application run for about 10 to 20 seconds so that the trace buffer collects runtime activity.

4. **Extract the trace buffer**

   Pause execution in the debugger.

   Extract the trace buffer.

   Open the Debug Console in VS Code.

   Enter the following GDB command to extract the trace buffer from RAM:

```
-exec dump binary value build/trace.bin *RecorderDataPtr
```

This command reads the internal trace buffer and saves it into a binary file called `trace.bin` under the build directory. You should not get any output after the command, if you get any warning or an error, that probably means that it was not configured correctly.

5. **Analyze the trace**

   Open **Percepio View** for Zephyr.

   Click **Recording Settings** on the right side or navigate to:

   File -> Settings

   From the left menu, select **Definition File Paths**.

   Percepio View requires a definition file generated by the build system. The file is typically named: `syscall-vVERSION.xml`.

   It is located inside the build directory:

```
<absolute/path>/my_philosophers/build/primary
```

   Add the full absolute path to the "Definition File Paths" section.

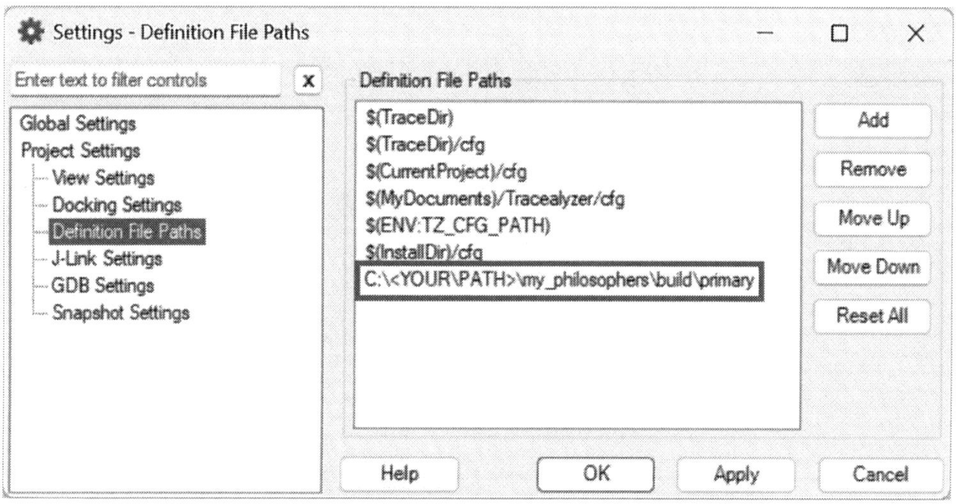

Figure 7.4 – Setting the build directory as a definition file path

   Once the definition file path is configured, open the recorded trace:

   File -> Open File

   Select the file `trace.bin` that you extracted earlier.

After loading the file, the trace viewer will display the execution timeline of your application.

You should see the different philosopher threads running, context switches between them, and the time spent in each state.

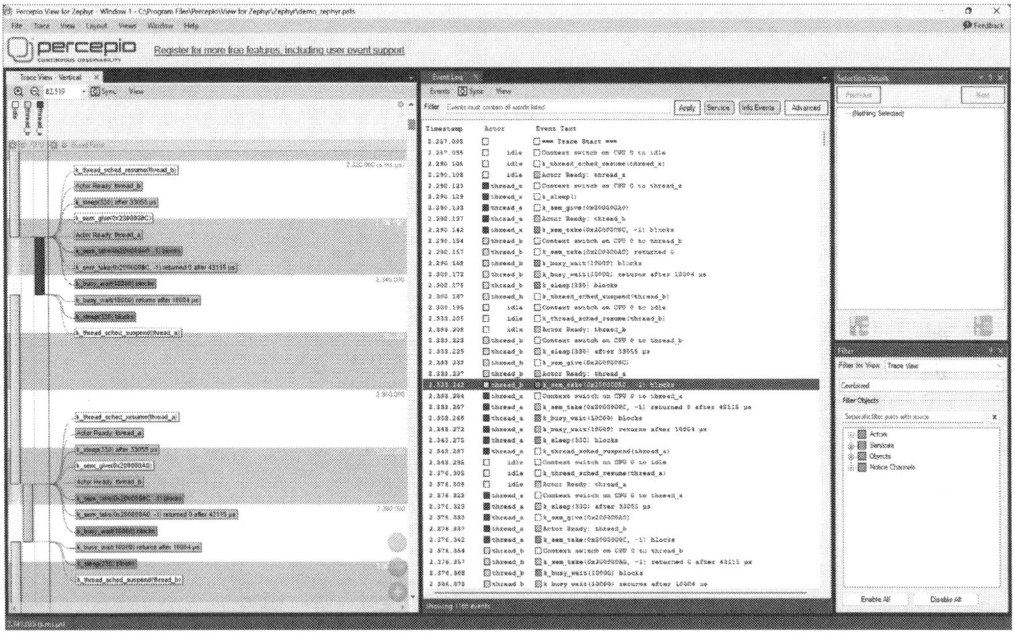

Figure 7.5 – Visualizing thread execution and context switches

How it works...

In this configuration, tracing operates in snapshot mode using a RAM ring buffer. The TraceRecorder library records runtime events such as thread scheduling, interrupts, and context switches into this buffer while the system runs.

Each trace event is stored sequentially in the buffer until it becomes full. Because the buffer operates as a ring buffer, older events may be overwritten by newer ones once the buffer reaches its capacity.

After the system runs for some time, the buffer contains a snapshot of recent system activity. This buffer is extracted directly from RAM using a GDB command and saved as a file (`trace.bin`).

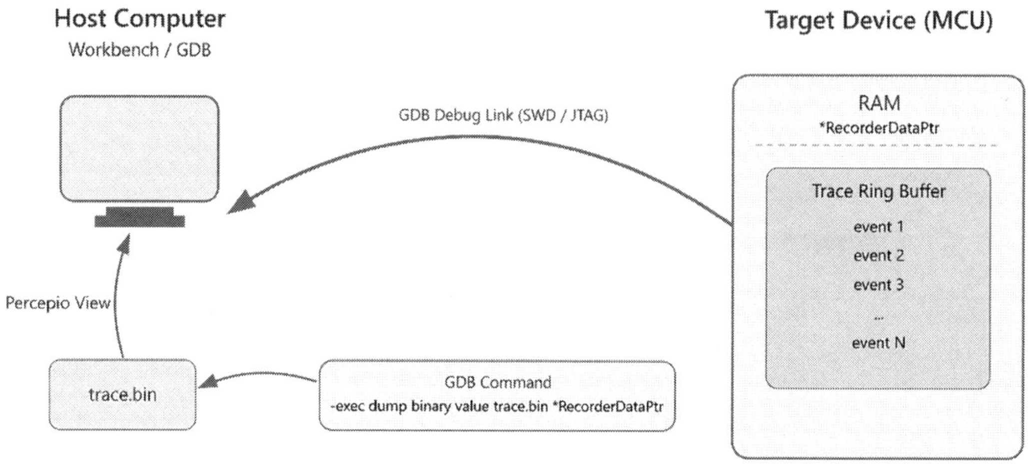

Figure 7.6 – Extracting a trace buffer from RAM for offline analysis

The trace viewer then parses this file and reconstructs the runtime behavior of the system.

Internally, tracing is implemented using tracing hook macros provided by the Zephyr tracing subsystem. These macros are normally empty, which means they generate no code and have no runtime cost when tracing is disabled.

When the TraceRecorder library is enabled, it defines implementations of these macros and records the trace information whenever they are invoked by the kernel.

Because tracing is implemented through conditional hook macros, the runtime overhead is minimal and completely removed when tracing is disabled.

There's more...

Registering Percepio View unlocks additional analysis features. For example, you can visualize the CPU Load Graph, which shows the percentage of processor time used by each thread and interrupt handler over time. This helps identify performance hotspots and periods of high processor utilization.

Another useful feature is User Events. These allow you to log application-specific events directly into the trace with minimal overhead. In the visualization, these events appear as markers in the timeline and are also listed in the event log.

If you use the full Tracealyzer tool, additional tracing options become available. For example, Tracealyzer supports other streaming backends such as SEGGER RTT and provides more advanced analysis views including communication flow graphs and memory usage statistics.

See also

- The instrumentation subsystem: `Zephyr Project Documentation, Tracing:` `https://docs.zephyrproject.org/latest/services/tracing/index.html`
- Percepio View for Zephyr: `https://traceviewer.io/`
- Tracealyzer: `https://percepio.com/tracealyzer/`
- SEGGER SystemView: `https://www.segger.com/products/development-tools/systemview/`

Get this book's PDF copy, code bundle, and more

Scan the QR code (or go to `packtpub.com/unlock`). Search for this book by name, confirm the edition, and then follow the steps on the page.

UNLOCK NOW

Note: Have your invoice handy. Purchases made directly from the Packt website don't require an invoice.

8

User Mode in Zephyr

Embedded applications often run with full privileges by default, meaning that any bug in one part of the system can affect the entire application. As systems grow in complexity or begin to include less trusted components, this lack of isolation becomes a serious risk, leading to memory corruption, unintended interactions, and difficult-to-debug failures.

Zephyr addresses this problem with **User Mode**, which allows selected threads to run with reduced privileges and controlled access to memory and kernel resources. However, enabling User Mode is not just a configuration change. It affects how memory is organized and how kernel objects are accessed. Without understanding these constraints, it can be difficult to design applications that are both secure and maintainable.

In this chapter, you will explore how Zephyr's User Mode works and how to design applications around it. You will learn how to enable `userspace` support, isolate global variables using memory domains, structure applications into logical components, and control access to kernel objects from user threads. Together, these recipes provide a practical approach to building safer and more robust Zephyr applications using User Mode.

In this chapter, we are going to cover the following main topics:

- Enabling User Mode (**userspace**)
- Isolating global variables with memory domains
- Designing applications with User Mode
- Accessing **kernel objects** from user threads

Technical requirements

To follow the recipes in this chapter, you will need:

- A supported host machine
- Zephyr SDK or a compatible toolchain
- A supported development board (e.g., STM32, NXP) or a simulation environment
- The target board should support memory protection hardware (**MPU/MMU**), a feature not available on all Cortex-M devices
- Visual Studio Code with Workbench for Zephyr extension (optional but recommended)

The code files for this chapter are available in the downloadable code bundle under the Chapter 8 folder. You can download the code by following the instructions in the **Download the example code files** section in the *Preface*.

Enabling User Mode (userspace)

In many embedded systems, all threads run with full privileges by default. This means that a bug in one thread can corrupt shared memory, misuse kernel services, or affect the rest of the system. User Mode solves this problem by allowing selected threads to run with restricted privileges and controlled access to memory and kernel objects.

In this recipe, you will enable `userspace` support, create a user thread with K_USER, and verify the difference between user and supervisor execution contexts at runtime. This mechanism prevents unintended memory corruption and protects critical kernel resources from faulty or untrusted code.

Getting ready

You need:

- A working Zephyr installation (west workspace set up)
- We will use the Zephyr Hello World sample as the starting point
- A board with hardware memory protection support (MPU or **PMP**)
- Serial console output working

If you are unsure whether your board supports User Mode, check your board's documentation for MPU or PMP support, or search for `CONFIG_ARCH_HAS_USERSPACE` in the Zephyr board configuration for your target.

How to do it...

1. **Create a new application from Hello World**

   Open the Create New Application panel in Workbench for Zephyr.

   - Select your target board
   - Select Sample project: `Hello World`
   - In Project Name, enter: `user_toto`
   - Finish the wizard

   This is functionally equivalent to copying the sample manually:

```
cp -r zephyr/samples/hello_world user_toto
```

   But using the Workbench panel ensures the project is properly registered in your workspace.

2. **Enable userspace in** `prj.conf`

   Open `prj.conf`, and add the following line:

```
CONFIG_USERSPACE=y
```

   This single symbol activates the entire userspace subsystem. Zephyr will enforce privilege separation and enable the memory protection hardware at boot.

3. **Create a user thread**

   Place the following code in `src/main.c`. This code creates a thread using `K_USER` and prints the value of `k_is_user_context()` from inside that thread:

```
#include <zephyr/kernel.h>

K_THREAD_STACK_DEFINE(u_stack, 1024);
static struct k_thread u_thread;

static void u_fn(void *a, void *b, void *c)
{
    ARG_UNUSED(a);
    ARG_UNUSED(b);
    ARG_UNUSED(c);

    printk("user ctx: %d\n", k_is_user_context());
```

```
    while (1) {
        k_sleep(K_SECONDS(1));
    }
}

int main(void)
{
    printk("userspace enabled? %d\n", IS_ENABLED(CONFIG_USERSPACE));
    printk("main ctx: %d\n", k_is_user_context());

    k_thread_create(
        &u_thread,
        u_stack,
        K_THREAD_STACK_SIZEOF(u_stack),
        u_fn,
        NULL,
        NULL,
        NULL,
        5,
        K_USER,
        K_NO_WAIT
    );

    return 0;
}
```

Note

`k_is_user_context()` returns 1 when the calling thread is in user context and 0 when it is in supervisor context. The `K_USER` flag passed to `k_thread_create()` tells Zephyr to start the new thread in user context instead of supervisor context.

4. **Build, flash, and run the application**

   Open the serial terminal. You should see output similar to:

```
userspace enabled? 1
main ctx: 0
user ctx: 1
```

   `main ctx:0` confirms that `main()` runs as a supervisor thread. `user ctx: 1` confirms that the thread created with `K_USER` is running in user context.

How it works...

When `CONFIG_USERSPACE=y` is set, Zephyr builds in full privilege-separation support. The hardware memory protection unit (MPU or PMP) is configured at boot, and the kernel gains the ability to run individual threads in an unprivileged CPU mode.

The main thread always starts in supervisor context. That is why `k_is_user_context()` returns 0 in `main()`. Supervisor threads are trusted: they can access any memory region and call any kernel function directly.

When you pass `K_USER` to `k_thread_create()`, Zephyr starts the new thread in unprivileged CPU mode. The hardware then restricts that thread to the memory regions it has been explicitly granted access to. Any access outside those regions raises a memory protection fault, which the kernel catches and handles typically by terminating the thread.

There's more...

When `CONFIG_USERSPACE` is enabled, you can control whether a newly created user thread inherits the kernel object permissions of its parent thread by using the `K_INHERIT_PERMS` flag.

If you create a thread like this:

```
k_thread_create(
    &u_thread,
    u_stack,
    K_THREAD_STACK_SIZEOF(u_stack),
    u_fn,
    NULL,
    NULL,
    NULL,
    5,
    K_USER | K_INHERIT_PERMS,
```

```
        K_NO_WAIT
    );
```

The new thread will inherit all kernel object permissions from the parent thread at the time of creation and permissions can be extended later as needed.

If `K_INHERIT_PERMS` is not specified, the new user thread starts without inherited kernel object permissions.

Using `K_INHERIT_PERMS` simplifies thread setup when a child thread should operate with the same kernel object access rights as its creator.

See also

- Zephyr Project Documentation, Threads: `https://docs.zephyrproject.org/latest/kernel/services/threads/index.html`
- Zephyr Project Documentation, Userspace : `https://docs.zephyrproject.org/latest/kernel/services/threads/index.html`
- RISC-V PMP and Zephyr RTOS - a new user(space) has entered: `https://baylibre.com/risc-v-pmp-and-zephyr-rtos-a-new-userspace-has-entered/`

Isolating global variables with memory domains

When multiple user threads share the same application, global variables can easily become unintended shared state. Without isolation, one thread may overwrite data that belongs to another component. Memory domains solve this problem by limiting each thread to selected memory partitions while still allowing controlled shared regions when needed.

In this recipe, you will place global variables into application memory partitions, assign a user thread to a memory domain, and observe what happens when the thread accesses memory outside its allowed domain.

Getting ready

Use the existing `user_toto` application.

How to do it...

1. **Define partitions and tag global variables**

   Open `src/main.c`, and add the following declarations at the top of the file:

   ```
   #include <zephyr/kernel.h>
   #include <zephyr/app_memory/app_memdomain.h>
   ```

```
#include <zephyr/sys/libc-hooks.h>

K_APPMEM_PARTITION_DEFINE(part_a);
K_APPMEM_PARTITION_DEFINE(part_b);
K_APPMEM_PARTITION_DEFINE(part_shared);

K_APP_DMEM(part_a) int a_var = 11;
K_APP_DMEM(part_b) int b_var = 22;
K_APP_DMEM(part_shared) int s_var = 99;
```

This creates three partitions. Each global variable is placed inside one partition at link time. `a_var` belongs to `part_a`, `b_var` belongs to `part_b` and `s_var` belongs to `part_shared`.

2. **Define a user thread that reads its allowed global variables**
   Add the thread function below the variable declarations. The access to `b_var` is commented out for now:

```
static void app_a_thread(void *p1, void *p2, void *p3)
{
    ARG_UNUSED(p1);
    ARG_UNUSED(p2);
    ARG_UNUSED(p3);

    printk("a=%d s=%d\n", a_var, s_var);

    /* printk("b=%d\n", b_var); */

    while (1) {
        k_sleep(K_SECONDS(1));
    }
}
```

3. **Define the domain and the partitions**
   Create a domain for "App A" that includes `part_a` and `part_shared`:

```
static struct k_mem_domain domain_a;

static struct k_mem_partition *domain_a_parts[] = {
#ifdef CONFIG_CPU_CORTEX_M
    &z_libc_partition,
```

```
#endif
    &part_a,
    &part_shared
};
```

> **Note**
>
> On some Cortex-M targets, include `&z_libc_partition` to allow user threads to call libc-backed functions like `printf`.

4. **Create the user thread and assign it to the domain**
   Define a user thread and then, in the `main()` callback, initialize the domain and assign the thread to it:

```
K_THREAD_DEFINE(a_tid, 512, app_a_thread,
                NULL, NULL, NULL, 5, K_USER, 0);

int main(void)
{
    k_mem_domain_init(&domain_a,
                      ARRAY_SIZE(domain_a_parts),
                      domain_a_parts);

    k_mem_domain_add_thread(&domain_a, a_tid);

    return 0;
}
```

5. **Build and run the application**
   Run the application and confirm you see output similar to:

```
a=11 s=99
```

6. **Verify memory isolation**
   Uncomment the line that accesses `b_var` and rebuild:

When isolation is working the user thread should trigger a fault when trying to access memory that is not in its domain

```
***** MPU FAULT *****
  Data Access Violation
  ...
```

How it works...

A memory domain is a set of memory partitions, and each partition represents a contiguous RAM region that the MPU/MMU can protect, as shown in the *figure 8.1*.

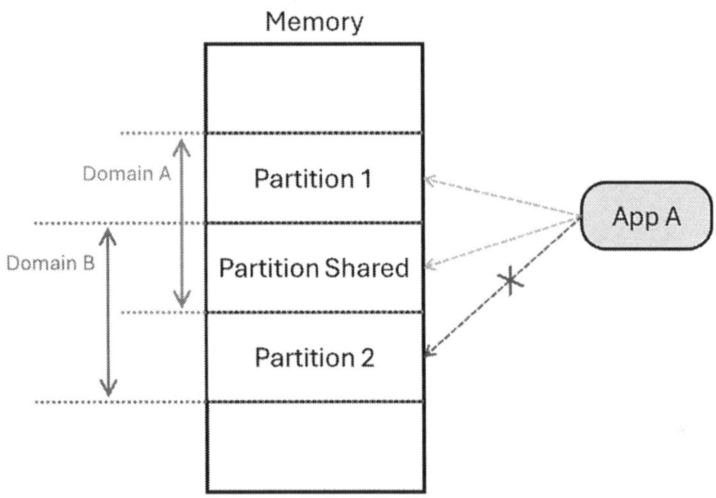

Figure 8.1 – Memory domains isolate application data

With automatic partitions, Zephyr's build system creates the partition regions, and `K_APP_DMEM(partition)` places the variable into that region. At runtime, once the thread is added to `domain_a`, it can access only the partitions included in that domain (in addition to its normal userspace memory such as its own stack and read-only program data).

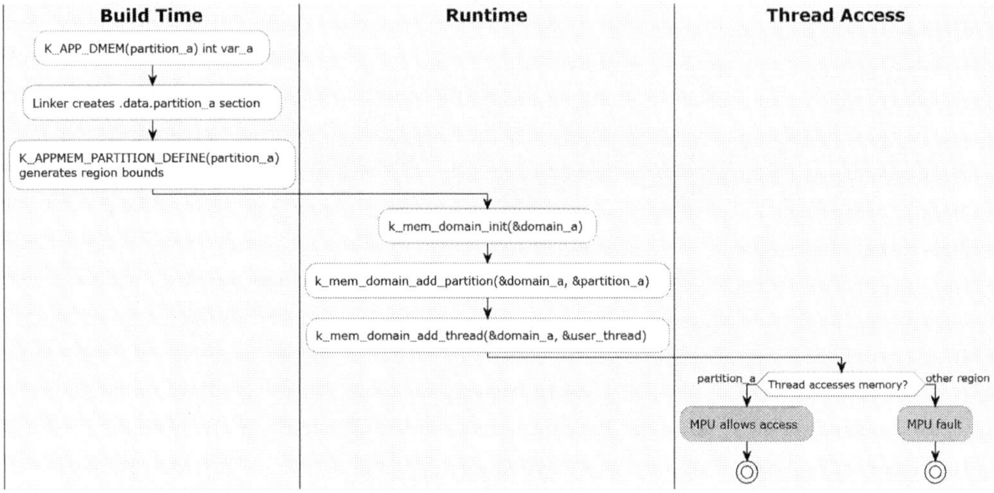

Figure 8.2 – Automatic memory partitions: build-time placement with K_APP_DMEM() and runtime access enforcement via memory domains and MPU

There's more...

In this recipe, you used automatic partitions with `K_APPMEM_PARTITION_DEFINE()` and `K_APP_DMEM()`. This is the most common and recommended approach because the build system handles memory placement automatically and keeps the application structure clean and scalable.

It is also possible to define memory partitions manually, where the memory boundaries and access permissions are specified explicitly. Manual partitioning may be required when working with special memory regions, external RAM, device memory, or when fine-grained control over access permissions is necessary.

For most applications, automatic partitions are sufficient and easier to maintain. Manual configuration is typically reserved for advanced designs or hardware-specific requirements.

See also

- Zephyr Project Documentation, Userspace Memory Domains: `https://docs.zephyrproject.org/latest/kernel/usermode/memory_domain.html`

- Zephyr Project Documentation, Application Memory Partitions: https://docs.zephyrproject.org/latest/kernel/usermode/memory_domain.html#application-memory-partitions
- User mode explained in simple words (Part 1): https://www.zephyrproject.org/user-mode-explained-in-simple-words/

Designing applications with User Mode

As applications grow, placing all user threads in a single memory domain becomes difficult to maintain. Independent components need isolated data, while some information must still be shared safely. Structuring the application into logical apps, each one packaged as its own Zephyr module, solves this problem.

In this recipe, you will split the project into logical apps, where each app lives in its own folder and is built as a Zephyr module. You will assign each app to its own memory partition, and connect user threads to the appropriate memory domains.

Getting ready

You need:

- The previous recipes in this chapter are completed
- Serial console output enabled

You can start this recipe in one of two ways:

- Continue directly from the `user_toto` application previously created.
- Create a new `Hello World` application and rename it, for example `user_apps_toto`, you need also to enable `CONFIG_USERSPACE=y` in `prj.conf`

How to do it...

1. **Create the project structure**
   Create the following directory layout:

```
user_apps_toto/
├── CMakeLists.txt
├── prj.conf
├── src/main.c
└── myapps/
    ├── app_a/
    │   ├── zephyr/
    │   │   └── module.yml
```

```
│   ├── CMakeLists.txt
│   └── app_a.c
├── app_b/
│   └── (same as app_a/, renamed for app_b)
└── shared/
    └── (same as app_a/, renamed for shared.c)
```

2. **Register the logical apps as Zephyr modules**
   In the top-level `user_apps_toto/CMakeLists.txt`, list the three app folders in `EXTRA_ZEPHYR_MODULES` before the `find_package(Zephyr ...)` line:

```
set(EXTRA_ZEPHYR_MODULES
    "${CMAKE_CURRENT_SOURCE_DIR}/myapps/shared"
    "${CMAKE_CURRENT_SOURCE_DIR}/myapps/app_a"
    "${CMAKE_CURRENT_SOURCE_DIR}/myapps/app_b"
)

find_package(Zephyr REQUIRED HINTS $ENV{ZEPHYR_BASE})
project(user_apps_toto)
```

   For Zephyr to treat each folder as a module, the folder must contain a `zephyr/module.yml` file.

   Add a minimal `module.yml` to each app instead, for example, for `app_a`:

```
build:
    cmake: .
```

3. **Assign each module's global variables to the right partition**
   In each module's `CMakeLists.txt`, place the module's global variables into a userspace partition with `zephyr_library_app_memory(<partition>)`.

   In `main.c`, add that partition to the correct memory domain for the threads that need it.

   Example: `modules/app_a/CMakeLists.txt`

```
zephyr_library()
zephyr_library_sources(app_a.c)
zephyr_library_app_memory(part_a)
```

`zephyr_library_app_memory(partition_name)` tells the build system to place all global variables defined in this library into the named partition. The partition itself is declared in `main.c` using `K_APPMEM_PARTITION_DEFINE()`.

4. **Define the entry function and data for each logical app**
   In modules `/app_X/app_X.c` (replace `X` with `a` or `b`):

```
#include <zephyr/kernel.h>

int app_X_cnt = 0;
extern int shared_value;

void app_X_thread(void *a, void *b, void *c)
{
    ARG_UNUSED(a);
    ARG_UNUSED(b);
    ARG_UNUSED(c);

    while (1) {
        printf("app_X %d, shared %d\n", app_X_cnt++, shared_value);
        k_sleep(K_SECONDS(1));
    }
}

K_THREAD_DEFINE(
    X_tid,
    1024,
    app_X_thread,
    NULL,
    NULL,
    NULL,
    10,
    K_USER,
    0
);
```

   In `shared/shared.c`:

```
int shared_value = 99;
```

In `shared/CMakeLists.txt`:

```
zephyr_library()
zephyr_library_sources(shared.c)
zephyr_library_app_memory(part_shared)
```

5. **Create partitions and domains and assign them to user threads**
   Similar to the previous recipes, we need the definitions of the partitions and domains in `src/main.c`, the domains must be added to the corresponding user thread:

```
K_APPMEM_PARTITION_DEFINE(part_a);
K_APPMEM_PARTITION_DEFINE(part_b);
K_APPMEM_PARTITION_DEFINE(part_shared);

static struct k_mem_domain dom_a;
static struct k_mem_domain dom_b;

static struct k_mem_partition *a_parts[] = {
#ifdef CONFIG_CPU_CORTEX_M
    &z_libc_partition,
#endif
    &part_a,
    &part_shared,
};

static struct k_mem_partition *b_parts[] = {
#ifdef CONFIG_CPU_CORTEX_M
    &z_libc_partition,
#endif
    &part_b,
    &part_shared
};
```

Define the memory domains in `main()` and attach threads to them:

```
extern k_tid_t a_tid, b_tid;
k_mem_domain_init(&dom_a, ARRAY_SIZE(a_parts), a_parts);
k_mem_domain_init(&dom_b, ARRAY_SIZE(b_parts), b_parts);
```

```
k_mem_domain_add_thread(&dom_a, a_tid);
k_mem_domain_add_thread(&dom_b, b_tid);
```

Build and flash, the expected output is similar to the following (with the counter incremented every second):

```
...
app_a 7, shared 99
app_b 7, shared 99
...
```

How it works...

Each logical application is assigned its own private memory partition for global data. This keeps internal state isolated from other applications while still allowing selected data to be exchanged through a separate shared partition when required.

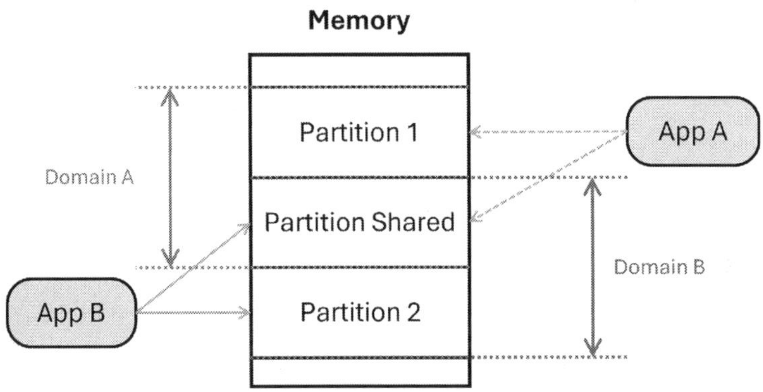

Figure 8.3 – Memory domains and shared partitions for logical apps

Each logical app is structured as a Zephyr module. The `EXTRA_ZEPHYR_MODULES` list, together with the `zephyr/module.yml` file in each app folder, tells the build system to compile that app and pull its `CMakeLists.txt` into the build. Inside that module, `zephyr_library_app_memory()` places the app's global variables into its dedicated partition. These partitions are then added to specific memory domains and assigned to the corresponding user threads.

As a result, each logical application can access its own private data and any explicitly shared data, but cannot access the private memory of other applications. By combining memory domains with automatic partition placement, applications remain modular, scalable, and easier to maintain.

`zephyr_library_app_memory(partition)` places an app's global variables into a dedicated partition at link time. A memory domain is a list of partitions a user thread is allowed to access. When you attach a thread to a domain with `k_mem_domain_add_thread()`, Zephyr applies that memory policy to the thread.

There's more...

Because each app is already a self-contained Zephyr module, moving one into its own repository needs no code change, only a change in how the build locates it. With west, instead of listing paths in `EXTRA_ZEPHYR_MODULES`, you add each app's repository as a project in your west manifest and run west update. West clones it and, because it contains a `zephyr/module.yml`, includes its `CMakeLists.txt` and `Kconfig` automatically.

Each logical app can then be developed, versioned, and tested independently, and reused across applications, while memory domains still enforce the same runtime isolation.

See also

- User Mode explained in simple words (Part 2): `https://zephyrproject.org/user-mode-explained-in-simple-words-part-2/`

Accessing kernel objects from user threads

In User Mode, user threads cannot automatically use **kernel objects** such as **semaphores**, mutexes, message queues, or devices. Access must be explicitly granted before a user thread can call kernel APIs on those objects. This protects critical system resources and prevents untrusted code from interfering with objects it should not control.

In this recipe, you will create a semaphore, start two user threads that both attempt to use it, grant access to only one thread, and compare the permitted and denied behaviors.

Getting ready

You need:

- Create a new Hello World application and rename it `user_syscall_toto`
- `CONFIG_USERSPACE=y` enabled in `prj.conf`
- Serial console output enabled

How to do it...

1. **Define the semaphore and thread resources**

   Open `src/main.c` and add the following declarations:

```
#include <zephyr/kernel.h>
K_SEM_DEFINE(g_sem, 0, 1);
```

2. **Add the user thread entry functions**

   Both user threads attempt to take the same semaphore:

```
void a_thread(void *p1, void *p2, void *p3)
{
    printk("A ctx: %d\n", k_is_user_context());
    printk("A take: %d\n", k_sem_take(&g_sem, K_SECONDS(1)));
}

K_THREAD_DEFINE(
    a_tid,
    1024,
    a_thread,
    NULL,
    NULL,
    NULL,
    5,
    K_USER,
    0
);

void b_thread(void *p1, void *p2, void *p3)
{
    printk("B ctx: %d\n", k_is_user_context());
    printk("B take: %d\n", k_sem_take(&g_sem, K_SECONDS(1)));
}

K_THREAD_DEFINE(
    b_tid,
    1024,
    b_thread,
    NULL,
    NULL,
```

```
        NULL,
        6,
        K_USER,
        0
    );
```

3. **Create both threads, grant access to only one, then give the semaphore**
   Implement `main()` to grant the semaphore to Thread `A` only, and then make the semaphore available:

```
int main(void)
{
    k_object_access_grant(&g_sem, a_tid);

    k_sleep(K_MSEC(200));

    k_sem_give(&g_sem);
}
```

   Thread B is started but never granted access to `g_sem`. The 200 ms delay before `k_sem_give()` gives both threads time to reach their `k_sem_take()` call before the semaphore becomes available.

4. **Run and validate**
   Typical output pattern:

   - Thread A runs first, and it can use the semaphore:

```
A ctx: 1
A take: 0
```

   - Thread B is denied:

```
B ctx: 1
...
Faulting instruction address (r15/pc): 0x69bcac14
```

How it works...

In User Mode, kernel objects such as semaphores, mutexes, queues, and devices are treated as protected resources. A user thread can use them only after a kernel thread grants explicit access.

When a user thread calls a protected kernel function, Zephyr cannot directly execute privileged code. The request must cross the user-kernel boundary and be checked before the kernel performs the operation.

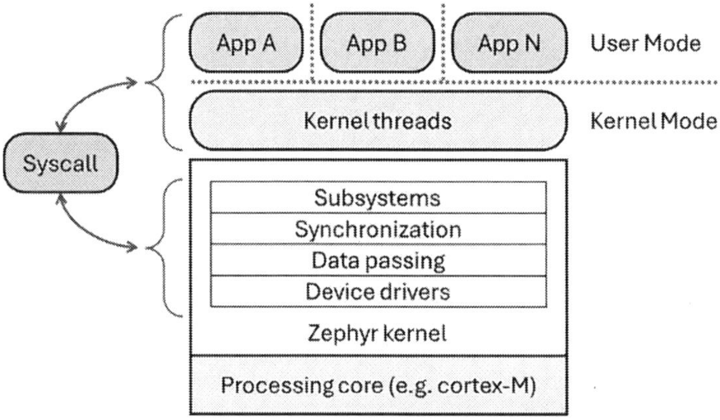

Figure 8.4 – User Mode architecture and the syscall boundary

When a `syscall` is made from User Mode, Zephyr routes it through three distinct layers before any privileged code runs. Think of it as a pipeline: the call enters at the top, gets checked at each stage, and only reaches the real implementation if everything looks correct.

The *figure 8.5* shows this pipeline:

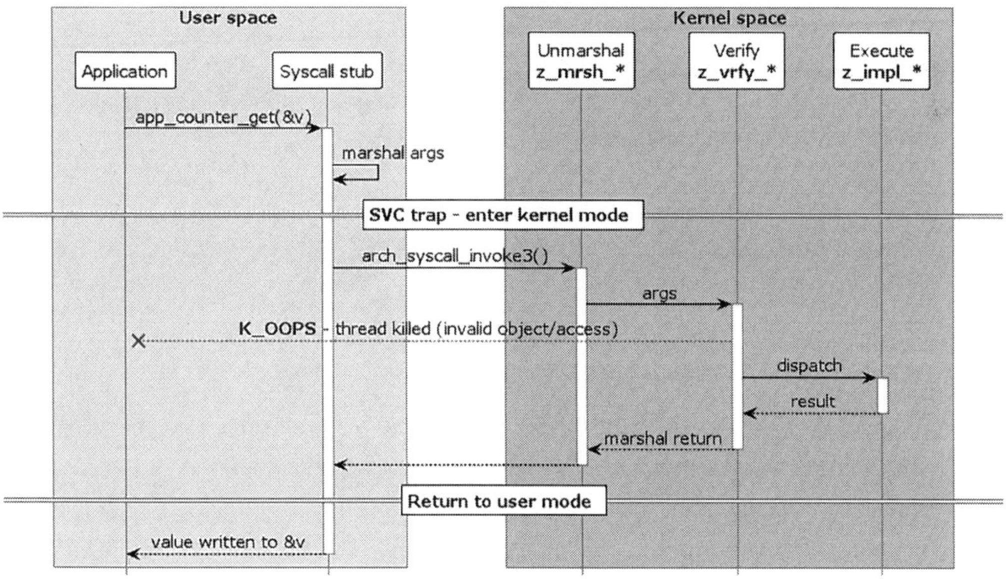

Figure 8.5 – Syscall path from user thread to kernel implementation

The first layer, z_mrsh_*(), packages the call arguments so they can be safely passed across the user/kernel boundary. The second layer, z_vrfy_*(), validates those arguments and checks that the calling thread actually has permission to use the requested object. Only if both checks pass does control reach z_impl_*(), which runs the real kernel implementation.

This layering is also practical for debugging. If a call fails because of a bad pointer or a missing permission grant, it will fail in z_vrfy_*() long before any kernel state is touched. If the arguments are valid but the behavior is wrong, the issue is in z_impl_*().

There's more...

Kernel objects in Zephyr are not limited to communication and synchronization primitives such as semaphores, mutexes, or message queues. Many other system resources are also implemented as kernel objects, including device drivers.

A user thread must be granted permission before it can interact with any of these objects. The complete list of kernel object types can be found in:

```
build/zephyr/include/zephyr/kobj-types-enum.h
```

Access to kernel objects can also be revoked at runtime. If a user thread should no longer use a specific object, a supervisor thread can call:

```
k_object_access_revoke(&g_sem, a_tid);
```

After revocation, any further attempt by that thread to use the object will result in a permission fault. This allows dynamic security policies, such as granting temporary access during initialization and removing it afterward.

See also

- Zephyr Project Documentation, System Calls: https://docs.zephyrproject.org/latest/kernel/usermode/syscalls.html
- Zephyr Project Documentation, Kernel Objects: https://docs.zephyrproject.org/latest/kernel/services/other/kernelobjects.html
- Zephyr Project Documentation, Semaphores: https://docs.zephyrproject.org/latest/kernel/services/synchronization/semaphores.html
- User Mode explained in simple words (Part 3): https://www.zephyrproject.org/user-mode-explained-in-simple-words-part-3/

Get this book's PDF copy, code bundle, and more

Scan the QR code (or go to `packtpub.com/unlock`). Search for this book by name, confirm the edition, and then follow the steps on the page.

UNLOCK NOW

Note: Have your invoice handy. Purchases made directly from the Packt website don't require an invoice.

9

Unlock Your Exclusive Benefits

Your copy of this book includes the following exclusive benefits:

Complete Code Bundle

Download the full source code for this book's examples and projects.

DRM-Free PDF Version

Download DRM-free PDF and ePub copies of this book.

7-Day Packt Library Access

Get 7-day unlimited access to 8,000+ books and videos. No credit card required.

Available for first-time Packt+ trial users only.

Next-Gen Reader Access

Read this book on Packt Reader with progress sync, dark mode and note-taking.

Follow the guide below to unlock them. The process takes only a few minutes and needs to be completed once.

Unlock this Book's Free Benefits in 3 Easy Steps

Step 1

Keep your purchase invoice ready for *Step 3*. If you have a physical copy, scan it using your phone and save it as a PDF, JPG, or PNG.

For more help on finding your invoice, visit `https://www.packtpub.com/en-us/unlock?step=1`.

> Note
>
> **Note**: If you bought this book directly from Packt, no invoice is required. After *Step 2*, you can access your exclusive content right away.

Step 2

Scan the QR code or go to `packtpub.com/unlock`.

On the page that opens (similar to *Figure 9.1* on desktop), search for this book by name and select the correct edition.

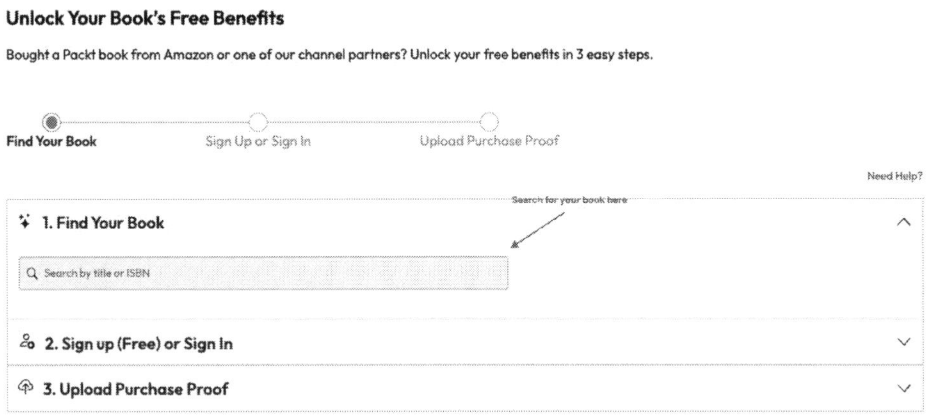

Figure 9.1: Packt unlock landing page on desktop

Step 3

After selecting your book, sign in to your Packt account or create one for free. Then upload your invoice (PDF, PNG, or JPG, up to 10 MB). Follow the on-screen instructions to finish the process.

Need Help

If you get stuck and need help, visit `https://www.packtpub.com/unlock-benefits/help` for a detailed FAQ on how to find your invoices and more. This QR code will take you to the help page.

> Note
>
> **Note**: If you are still facing issues, reach out to `customercare@packt.com`.

Other Books You May Enjoy

If you enjoyed this book, you may be interested in these other books by Packt:

Hands-On RTOS with Microcontrollers

Jim Yuill, Penn Linder

ISBN: 9781803237725

- Understand RTOS use cases, and decide when (and when not) to use real-time OS
- Use the FreeRTOS scheduler to create, start, and monitor task states
- Improve task signaling and communication using queues, semaphores, and mutexes
- Streamline task data transfer with queues and notifications
- Upgrade peripheral communication via UART, USB, and DMA by using drivers and ISRs
- Enhance interface architecture with a command queue for optimized system control
- Maximize FreeRTOS memory management with trade-off insights

Packt is searching for authors like you

If you're interested in becoming an author for Packt, please visit `authors.packt.com` and apply today. We have worked with thousands of developers and tech professionals, just like you, to help them share their insight with the global tech community. You can make a general application, apply for a specific hot topic that we are recruiting an author for, or submit your own idea.

Share your thoughts

Now that you've finished *Zephyr RTOS Cookbook*, we'd love to hear your thoughts! Scan the QR code below to go straight to the Amazon review page for this book and share your feedback or leave a review on the site that you purchased it from.

https://packt.link/r/1807429172

Your review is important to us and the tech community and will help us make sure we're delivering excellent quality content.

Index

Made in the USA
Coppell, TX
08 October 2026

92261505R00125